# THE WORLD ITSELF

Northville District Library
212 West Cady St.
Northville, MI 48167

# THE WORLD ITSELF

## Consciousness and the Everything
## of Physics

## ULF DANIELSSON

BELLEVUE LITERARY PRESS
NEW YORK

First published in the United States in 2023
by Bellevue Literary Press, New York

For information, contact:
Bellevue Literary Press
90 Broad Street
Suite 2100
New York, NY 10004
www.blpress.org

© 2020 by Ulf Danielsson
Originally published by Fri Tanke förlag in Swedish as *Världen själv*
Translation copyright © 2023 by Ulf Danielsson

Library of Congress Cataloging-in-Publication Data

Names: Danielsson, Ulf H., author.
Title: The world itself : consciousness and the everything of physics /
    Ulf Danielsson.
Other titles: Världen själv. English
Description: First edition. | New York : Bellevue Literary Press, 2023. |
    "Originally published by Fri Tanke förlag in Swedish as Världen själ"--
    title page verso.
Identifiers: LCCN 2022021757 | ISBN 9781954276116 (print ; acid-free paper) |
    ISBN 9781954276123 (ebook)
Subjects: LCSH: Physics--Popular works.
Classification: LCC QC24.5 .D3613 2023 | DDC 530--dc23/eng/20220805
LC record available at https://lccn.loc.gov/2022021757

All rights reserved. No part of this publication may be reproduced or
transmitted in any form or by any means, electronic or mechanical, including
photocopy, recording, or any information storage and retrieval system now
known or to be invented, without permission in writing from the publisher,
except by a reviewer who wishes to quote brief passages in connection with a
print, online, or broadcast review.

Bellevue Literary Press would like to thank all its generous donors—individuals
and foundations—for their support.

This publication is made possible by the New York State
Council on the Arts with the support of the Office of the
Governor and the New York State Legislature

Book design and composition by Mulberry Tree Press, Inc.

Bellevue Literary Press is committed to ecological stewardship in our
book production practices, working to reduce our impact on the natural
environment.

∞  This book is printed on acid-free paper.

Manufactured in the United States of America.

First Edition

10   9   8   7   6   5   4   3   2   1

paperback ISBN: 978-1-954276-11-6

ebook ISBN: 978-1-954276-12-3

# Contents

# THE WORLD ITSELF

# Reality *Is* Real!

N 1976, THE AUSTRIAN-BORN American psychologist, family therapist, and philosopher Paul Watzlawick published the book *How Real Is Real?* In it, he defends a thesis that is very dear to constructivist theories or, more broadly, postmodern theories—that is, we construct our own reality and, therefore, there is no objective reality. Somehow, these theories underlie the current "post-truth" world.

I must admit that, as a physicist, I have always found this denialism totally abnormal. Those who do not believe in the reality of a wall standing right in front of them could perhaps try banging their heads on it to see if they gain some sense of reality. The principle of action-reaction states that the force the head exerts on the wall is associated with another opposing force exerted by the wall on the head, which can damage biological tissues, possibly causing loss of consciousness. This would be a "reality shock." Obviously, the real world, including the wall, continues to exist, even when we have no perception of it. If the world did not really exist, and if the human community could not share fairly accurate descriptions of the world, not only would the existence of physics and other sciences be jeopardized but, more serious than that, the life of that community would be endangered.

The Swedish physicist Ulf Danielsson has, like me, no

doubt about the existence of reality. Our job as physicists is studying "the world as it is"—not as we would like it to be, but simply as it is. We live in a real world that overwhelmingly imposes itself on us, regardless of the enticing fantasies about alternative worlds we may construct. The world, which we also call the universe, is vast, varied, and complex. Perhaps the ultimate complexity can be found in the organization and functioning of our brains, which is the part of the world that tries to understand it (to do this, you should avoid banging your head against the wall). Through the scientific method, in physics and other sciences, we have managed to obtain representations of the world that seem true to us, in the sense that they fit relatively well with reality and are very useful to us because they allow us to live better. Generally, the world is a bleak and dangerous place, and only our knowledge of it can turn it into a livable and comfortable location.

An essential ingredient of the scientific method is mathematics: Physics describes the world through laws that have a mathematical expression. Galileo, the father of the method, said, using a beautiful metaphor, that the "Book of Nature" is written in the language of mathematics and that only those who understand that language will be able to read it. It was through mathematics that the Italian physicist described falling bodies. It was through mathematics that later Newton realized that an apple falling on his head (a much smaller shock than a head banging on a wall) was governed by a law, the law of universal gravitation, which also governs the movement of the Moon around the Earth and the Earth and the other planets around the Sun. That description contains not only simplicity but also beauty. According to this law, the force of attraction between two bodies is directly proportional to their masses and inversely proportional to the square of the distance between them. The other sciences have, some more and some less, followed this path of mathematization that physics was the first to adopt.

Danielsson is a theoretical physicist, professor at Uppsala University (the university where Linnaeus and Celsius taught), an expert in string theory and cosmology, which are subjects where sophisticated mathematics is absolutely necessary. In the quest to answer the big questions—What is the world made of? How did the world come into being? What will happen to the world?—physicists have achieved great success; although our knowledge is incomplete, we have discovered that behind the immense complexity of the world, there are simple rules— that is, there is a "hidden order." Not content with the partial knowledge of this order they have already obtained, physicists seek a "theory of everything," a unified theory of elementary particles and fundamental interactions (string theory is a candidate, but it has not passed the crucial test of experiment yet). With a good knowledge of the order of the world that physics has already revealed, Danielsson leaves us a very important message in his book: Please do not mistake the world, which is real, with our descriptions of the world, which are only human attempts to represent it, and which, as the whole history of science has taught us, can be improved. Reality is one thing, and the representations we make of it, particularly mathematically based representations called "laws of nature," are another. The world is as it is, and our images of it can be improved. Actually, they have been, as shown, for example, by the description of gravitation obtained by Einstein, which, although containing Newton's description, goes far beyond it. The Swedish physicist draws attention to the fact that, with Einstein, the concept of force was dispensed with, since the apple, or the Moon, or the Earth, performs its movements following only the curved geometry of space-time. Therefore, the concept of "force," although useful, was temporary.

Danielsson is very clear: The universe is more than the laws we discover; the universe is not mathematics. We cannot mistake reality with models, and our computer simulations of the world are nothing but "caricatures" of reality. He goes even further: Contrary to what we sometimes read and hear, the universe is not a computer. We ourselves and all living beings, although part of the physical world, are not machines. Contrary to what some artificial intelligence scholars maintain, the brain is not a computer.

The author is well aware that the concept of reality is elusive. It belongs to the domain of philosophy, so he does not hesitate to explain what philosophy has said about reality. He cites, among others, Aristotle, Husserl, and Heidegger. He explains that the error of attributing reality to our mental constructions goes back to ancient Greece, going by the names of Pythagoreanism and Platonism. He says, as Damasio had already said, that Descartes was mistaken in his theory of body-soul dualism: Descartes separated the body from the soul, but we know today—you just have to look at the continuum of biological species—that the mind (the new name for the soul) is inseparable from the body and that its results, however imaginative, can only be explained through the sensory experience of the body. Danielsson highlights that Descartes died in Sweden: This book of his sets out arguments for a "second death" of the French sage. In contemporary philosophy, the author elaborates on theories of consciousness. He does not deal with postmodern philosophies, whose nefarious consequences the reader will become aware of by reading *Cynical Theories,* by Pluckrose and Lindsay (Pitchstone Publishing, 2020).

Danielsson addresses profound questions that philosophy has long been dealing with in connection with science (the relationship used to be so close that physical science was called "natural philosophy"). This is the case of determinism and free will: Is everything determined as natural laws seem to indicate, or do we have free will? The author offers an ingenious solution: The notion of free will is tainted because it is proposed in contrast to the notion of determinism, which is nothing but a characteristic of some of our models of reality. There is really no free will, just as there is really no determinism.

The conclusion can only be that our physics is incomplete, very much a hostage of Platonic idealism, clearly evident in the equations of string theory. I would say that, by trying to get to the bottom of things, Danielsson saw that perhaps there is another way of looking at them. Maybe we need to look at the whole without being obsessed with the bottom. The equations we write lack reality. Or, in other words, they lack a body. Not knowing what the physics of the future will be, Danielsson states that physics will have much to learn from biology, which has unified, without much mathematics, the prodigious variety of the living world. Danielsson does not say so, but I dare say it: Maybe there is no theory of everything. Or, if we insist on giving in to the temptation to describe reality with a unified theory, perhaps the theory of everything is of a different kind.

This is the sixth book on popular science by this author, who, at fifty-eight, has extensive experience in disseminating science in the Swedish press as well as on radio and television (he has even performed at the Royal Dramatic

Theatre in Stockholm). After graduating from Uppsala University, Sweden's oldest university, he earned his Ph.D. at Princeton University, in the town where Einstein lived, under the guidance of a Nobel Prize winner in physics, the American David Gross. Danielsson has been a member of the Royal Swedish Academy of Sciences, which awards the Nobel Prizes, since 2009. At the announcement of the 2020 Nobel Prize in Physics, which distinguished the British Penrose, the German Genzel, and the American Guez for their work on black holes, it was Danielsson who explained, *urbi et orbi,* what they are: terrible abysses of the universe, where space and time bend brutally.

This book reads very well, thanks to the author's clarity of thought and the quality of his writing. It was proficiently translated from the Swedish original. Even if the reader has no background whatsoever in physics or science in general, he or she will be enticed without having to agree with everything. One of the secrets is Danielsson's use of literary references: He invokes Proust, Borges, and Rushdie. And he makes reference to the visual arts and cinema, mentioning Escher, Tarkovsky, and Kubrick.

I particularly liked how he includes examples from everyday life in his text. For example, a promising soccer player does not need to solve any equations to score a goal. He also punctuates the text with personal anecdotes. For example, when a teacher in the kindergarten his child attended asked him what infinity was, he replied, using his body, that when you go from here to there, you can try to go farther and farther and farther. You will obviously become tired at a given point. I mean, infinity is an ideal, platonic notion, which is corporally inaccessible to us. We do not know if the universe is finite or infinite, but the part that is accessible to us, whether via our physical movement or our artifacts or the reception of light signals, is finite. Knowledge about the universe, however, seems to have no end in sight, as this book so aptly suggests. Reality is real, and we are and will be faced with it.

*—Carlos Fiolhais,*
PROFESSOR OF PHYSICS AT
THE UNIVERSITY OF COIMBRA

# 1

## Everything Is Physics

*Natural objects . . . must be experienced before one can formulate theories about them.*

—EDMUND HUSSERL

I HAVE A SECRET TO TELL YOU: Living beings are not machines, there is no mathematics outside of our heads, the world exists and it is not a simulation, computers cannot think, your consciousness is not an illusion, and your will is not free.

I am a theoretical physicist and make a living exploring the foundations of the universe with the help of mathematics. History has shown that it is a successful method and has led to a comprehensive picture of everything we find in the universe. Physics has revealed how the world around us is composed of microscopic elements governed by universal laws, and how the universe has a history that stretches back almost fourteen billion years to a beginning. Intoxicated by success, we easily forget that, after all, there is a difference between the mathematical models and the real, physical world.

Mathematics does not rule the universe; we use it to describe what we discover in the universe. What applies to mathematics also applies to the laws of nature. There are no laws of nature out among the stars or in the innermost parts of the atom. They are just our way of summarizing our knowledge of the universe. Nature is what it is, while we as biological organisms try to understand as best we can what we experience.

These misunderstandings are based on a dualistic view of existence with historical roots, where human consciousness is raised above the world itself. We dream of an eternal and extraterrestrial sphere that is set to control mortal matter. Despite all that science has revealed about the universe, we have not succeeded in liberating ourselves from a worldview that is fundamentally religious. We continue to use concepts and metaphors that corrupt our thinking. Physics is presented as a science that discovers beautiful and mathematical laws with an independent and autonomous existence that rules over matter. Searching for the simple and beautiful has in many contexts been a successful methodology, but it also has its risks. There are no guarantees that the universe in any fundamental sense is either beautiful or simple.

All this is secretly and intimately connected with the belief in a transcendent soul. But that the self is rooted in the body must be true for the simple reason that mathematics, language, symbols, and, more important, meaning and semantics, do not exist without a physical body. The self itself is not an illusion: It is embodied and a property of matter that physics must be able to describe.

In the book that you hold in your hand, I argue that everything is physics and that there is no reality outside of matter. But there is no reason to believe that we are even close to understanding what this world of matter is capable of. Given the limitations set by biology, it is likely that there are central truths about the world that are so far beyond our comprehension that we cannot even formulate the questions. It is not about inaccessible and exotic phenomena, but aspects of the very same world that we find ourselves to be part of and experience in our daily life. To understand how subjective experience can exist, and the difference between living matter and dead machines, we will need a completely new approach. It is not about small details, but about paradigm shifts of the same dignity as when Newton formulated mechanics or when relativity and quantum mechanics were presented. It is also crucial for how we look at ourselves and how we live and value our lives.

## What the World Is

We live in a time when the line between imagination and reality is blurred. This is not just because of the role fiction plays in entertainment and social media but also because of our relationship to nature and the physical basis of our well-being and survival. We have imagined that we primarily live in a culturally and socially constructed world where we set the rules ourselves. In popular culture, people play with the idea that the entire universe can be equated to a simulation that is run on a computer. There are even those who have become so out of touch with the world that they seriously imagine that our consciousness, with the help of technology, can free itself from our bodies and be uploaded in electronic form. The way in which physics is presented is far from the sometimes quite brutal, and at the same time wonderful, truth about ourselves that science has revealed. We are biological beings trapped in our organic and mortal bodies but at the same time part of a universe that has never appeared more enigmatic and magnificent.

Even among those who claim to represent science, there are many who seem to doubt whether the world really exists. They are so seduced by mathematics and computer science that they see no difference between the real world and what can be simulated inside a computer. There are even those who believe that the world itself is basically pure mathematics. They are fascinated by what physics has to say about the universe and find the power and beauty of mathematics overwhelming. The result is a superstition of almost religious proportion about what good or evil technology can achieve. Many also conclude that the very computers that helped us make all these discoveries are the best metaphor for the universe itself. The phenomenon is not new. In the nineteenth century, the forefront of technology lay in steam engines, and the universe was likened to a mechanical marvel. Now we think we know better and believe that the universe can be likened to a computer. Or even better: The universe is nothing but pure information, where even you and everything you are, including your thoughts, can be translated into a series of ones and zeros. At best, maybe this could be a way to give you eternal life. Surely that would be wonderful.

The only problem is that it's not true.

All the things we can see and experience, our material universe as well as life and consciousness, are different aspects of the same world. The task of science is to find out as much as possible about how this, our universe, works. It does so with the help of various disciplines such as physics, chemistry, and biology, which through medicine and neuroscience translate into the humanities, social sciences, and philosophy. They all have their own language and their own criteria for truth. We see these different parts of knowledge as more or less independent of one another. There is a widespread belief that at the base of everything, or perhaps it is at the top, is physics, from which everything else can be derived. A biologist needs to know some chemistry to understand the processes that take place in the cells of an animal or in a plant. The chemist, on the other hand, trusts the physical theory of atoms. The discoveries of fundamental physics, filtered through layer after layer, are essentially irrelevant to most of us. Although the universe in all its complexity and beauty is assumed to be a consequence of fundamental laws in particle physics, this knowledge is not of much use to a chemist and even less so to an expert in Nordic mythology. Speculation in the field of quantum gravity is not something that an expert on how birds migrate needs to care about, and similarly, most physicists are convinced that none of what one can learn about living beings is relevant to the mathematical study of the big bang. Our universe, and especially our knowledge of it, is conveniently packaged to make our life as scientists as simple as possible. In this way, there is room for sciences other than physics, both natural sciences and social sciences. Of course, pioneering scientific advances are often made at the border between usually separate subject areas, where new perspectives on old problems are discovered. Despite this, the long-established and overarching hierarchy is hardly questioned.

I must admit that I have a view of the world that at first glance may seem even more extreme. It is not just that physics is the basis of everything; it is everything. I define physics as the study of the world itself in all its aspects. It is a world of which we as organic beings form a part, and through evolution we have slowly become aware of ourselves as matter awakened from its eternal slumber. Physics is not about how a free and independent observer floats outside the world and observes it from a proper distance. Our organic bodies, all our thoughts, including the scientific models we create, are parts of the same world that we so desperately want to grasp. The physics I imagine must handle everything; nothing must be left aside. It is literally a matter of life and death.

## The Origin of Dualism

In his remarkable *The Phenomenon of Life,* the German-born philosopher Hans Jonas (1903–1993) tells a story about how the world has changed through the millennia. It is about how life had to give up in the face of death. In ancient times, the world was full of life. There were people, animals, and plants of all kinds that spread over the Earth. Some were appreciated and praised; others were dangerous and hid in the dark and were sources of myths and fairy tales. Everyone strived in his own way to survive, so surely natural forces such as the wind and the sea, in constant motion, also had to be alive. Perhaps the celestial bodies also constituted life-forms that deliberately followed their paths with purposes unknown to us. To make everything understandable, the people of antiquity constructed cosmologies where life was taken for granted and death constituted the great mystery. What role could death play in a world so full of life? How could you understand it? There was a need for ideas that could point to a continuation beyond the end of life to deal with the contradiction between life and death.

Everything changed when science discovered that the world was so much larger than the early humans could have imagined. After realizing the true place of the Earth in the universe, it became clear that most of the universe consisted of dead matter. Death was everywhere, with life as the unusual exception. The worldview changed: Death became the rule and life the mystery in need of an explanation.

With this sharp contrast between life and death, the

worldview, according to Jonas, began to become dualistic. Dualism makes a distinction between body and soul, where the body consists of perishable matter, while the soul is spiritual and eternal. Most of the universe was certainly dead matter, but no one could deny the existence of the inner light, the self, so obvious to all of us, a living spirit that must have an origin beyond this world. Maybe this could somehow defeat death and live forever? This soul, with its ability to overcome death, must be something so much bigger than dead matter and be exactly what makes us humans unique. This was the idea. The living body was abandoned and left as dead matter, ingeniously constructed, of course, but basically nothing more than a machine controlled by the immortal and immaterial soul.

Through this approach, modern science became possible. The scientist remains separate from the dead world of matter and can study it from a safe, elevated, and objective perspective with sensitive instruments and sum up the results in the form of mathematical laws. These laws capture everything we can learn about the world, and in this sense, science became obsessed with death.

Jonas argues that this comfortable division of body and soul, a world of dead matter in contrast to a living human self, falls apart when the theory of evolution is formulated. The dualism that philosophers like René Descartes (1596–1650) advocated and that formed the very foundation of modern science was challenged by the continuum of life that evolution revealed. One could no longer easily argue that there was something that set us apart from other life on the planet. All these other beings, whether amoebas, butterflies, dogs, or chimpanzees, were, according to Cartesian dualism, nothing but machines made of dead matter. We humans, on the other hand, were assumed to be fundamentally different, with our souls, our selves, and our ability to study the world. Could this really make sense? When were these decisive and wonderful steps taken in evolution? Did *Homo erectus* already have an inner self, or did it arise later, perhaps when we acquired language?

Evolution occurs in steps, but there is a great deal of continuity. It is reasonable to believe that what defines us as human beings—whatever it is—may have existed in our ancestors in reduced form and even in other life-forms. How much, and to what extent, we do not know, but it is no longer possible to draw any clear line between consciousness and dead matter. One possibility, of course, would be to completely resign oneself to dead matter, and let it invade and conquer the last refuge of the living, the self, and let this be explained as nothing more than an illusion. The old world so full of life has then been fully replaced by a world of dead machines. Dualism is dead and, with that, life and the subjective self.

But, as we shall see, there is a threatening contradiction. The way we conduct science is still stuck in the dualistic framework. How can an objective vantage point for science exist if there are no vantage points at all? How can there be illusions in a world of only machines, where there is no one who sees this, no one to be deceived?

## Consciousness Is a Physical Phenomenon

The most amazing physical phenomena we know are life and consciousness. If everything is physics, there must be an unbroken line that can be drawn from the hot and dense plasma of the big bang to the human consciousness. Physics must be able to describe this. The inner self is as real as a thing or any other aspect of matter and must be woven without contradiction into our image of the world.

I do not agree with those who consider consciousness to be an illusion and that living organisms are best described as machines. I also do not agree with those who believe that consciousness must forever be inexplicable and that it is by definition outside of what is physics. Certainly consciousness is a difficult problem, even "the hard problem," to quote the Australian philosopher David Chalmers. But I do not believe that the self can be separated from its material basis, whether in the form of an immortal soul or its modern substitute: information. Both cases are wishful thinking and fairy tales.

Can physics really be everything? Some find such a worldview deprived of all meaning and value. Surely there must be aspects of the world that are not captured. Even in principle. This longing for something more is satisfied by religion for some, while others dream of how consciousness in the form of information can be transferred and uploaded to a computer and in this way give us eternal life. Some equate the world itself with pure mathematics. But what I call matter fundamentally involves life and consciousness—including the indispensable subjective first-person perspective that is constantly present. We cannot stand outside the world; we are in the midst of it as living bodies, and we always look at the world from the only vantage point that exists: from within. Physics—in the way I choose to define it—does not disenchant the universe, but reveals a reality far more wonderful than we could have imagined with our limited intellect.

In his breakthrough work, *This Life,* the philosopher Martin Hägglund describes how one of the fathers of the Christian Church, Augustine, is torn between his love of the perishable Earth and the allure of eternal heaven. The fundamental problem, Hägglund argues, is that the transient is meaningless in the perspective of eternity. It is only when something is at stake that there can be a meaning. If time is dissolved and nothing can happen anymore, you might just as well be dead. Heaven does not give the earthly things we love eternal life, but instead denies their significance. The science of our time faces a similar dividing line. Belief in heaven corresponds to the dream of a world that can be fully captured by the mathematical information in a computer program. Mixing model and reality in this way is basically an attempt to stand outside the world and to lift oneself out of the uncertain and perishable.

This is doomed to fail.

We are biological beings defined by the very creation of meaning for which perishable is a prerequisite. The crucial insight, contrary to everything we in the world of today are encouraged to believe, is that the world is real. There is a difference between what is only fantasy and what is of flesh and blood. That's what this book is about: the world itself.

# 2

## Living Beings Are Not Machines

*Why did Newton's apple fall to the ground? Because physics comes first here, I will start with the answer from the physicist.*

*"There is a force called gravity that exists between any two objects, with a magnitude proportional to the product of the objects' masses, and inversely proportional to the square of the distance between the centers of the masses. The apple is therefore propelled by this force towards the center of the earth when it leaves the tree."*

*Now we will move to the answer from the evolutionary biologist: "Once upon a time, apples used to go in all directions, up, down, sideways— the world was full of ricocheting apples. However, only those apples that fell to the ground were able to germinate and grow new trees."*

—Frances Arnold,
Nobel laureate in chemistry,
speech at the Nobel Banquet, 2018

WHEN I WAS SIX YEARS OLD, I wanted to become a biologist. I was passionately interested in animals of all kinds, especially the slightly more alien ones like fish, snakes, insects, and, of course, dinosaurs. I kept my interest up as I grew. I got an aquarium with tropical fish when I was nine and in summer I caught mysterious creatures in a nearby pond. I followed tadpoles that grew into frogs and how larvae that I fed with stinging nettles turned into beautiful butterflies. But as my interest in science matured, I felt that the deepest questions, and answers, could not be found among the living creatures I caught in the forests, meadows, and lakes of my childhood. If I were to understand how the world really worked, it was the starry sky that mattered.

So I chose to become a physicist. Physicists of my kind study the fundamental laws of nature and do not need to worry about biology. A biologist, on the other hand, must learn at least some basic physics. The experimental methods of physics in the form of microscopes and other instruments are, after all, absolutely indispensable to achieve anything in the biological sciences. But what can a physicist learn from a biologist?

I had the chance to discuss these issues with biologist and popular science writer Richard Dawkins when I interviewed him onstage in front of nearly a thousand people and he had nowhere to escape. He is the perfect British gentleman. He speaks softly and with humor and is in reality a completely different person from how he is sometimes portrayed in the media. For some he is a hero, while others fear him for his uncompromising crusade against religion.

As a physicist, I felt flattered by the way Dawkins described my subject. He believes that physics is the senior science, on which everything else is based. Biology deals with the complex, while physics takes care of the profound. In this I recognized my own reasons for choosing physics over biology. It was the profound that exerted attraction and I had no problem leaving the complex and dirty world of biology to others.

Physicists, at least those who are addicted to subjects such as particle physics, often are obsessed with simplicity and beauty. Even as they support Dawkins's atheism as it pertains to the biological sciences, when it comes to their own subject, they gladly profess a belief in the necessity that beautiful and simple laws form the basis of creation. Atheists or not, they jokingly talk about seeing God's face or about God not playing dice. Are these no more than metaphors? For many, sure, but if you take the position that there is beauty and simplicity at the root of everything, you quickly run into the question "Why?" Who ordered all the beauty of the natural laws that we take for granted? In whose eyes should it be beautiful? And on what scale should simplicity be judged? Why would it be so simple that we can understand it? If you believe that beauty is an absolute concept, independent of man, a tempting and almost inevitable conclusion is close at hand. Surely a creator must be hiding somewhere with a taste similar to ours and a mind not much bigger. Isaac Newton (1642–1727) and his contemporaries had no problem with this concept. On the contrary, they wanted to find evidence that God existed and confirm intelligent design.

In my discussion with Dawkins, I wanted to find out if he thought that physics has nothing left to learn from biology. This has not always been the case. There are important examples from history when clues to new physics were hidden in insights that had come from biology. Evolution is just such a case.

Lord Kelvin (1824–1907), a prominent figure in physics of the late nineteenth century, once claimed that physics had revealed all the secrets of nature and that only a few annoying little details remained. These details would lead to quantum mechanics as well as to relativity, but all that belonged to the future. William Thomson, Lord Kelvin, along with other physicists, had set a fixed limit for the age of the Earth of no more than a few tens of millions of years. The magma in the Earth's interior could not occur in molten form, and no energy source could have caused the Sun to shine, for a longer time than that. This, of course, posed a problem for Charles Darwin (1809–1882), who needed much longer periods of time than that for evolution to work. In a letter to another great biologist, Alfred Russel Wallace (1823–1913), he acknowledged that "Thomson's views on the recent age of the world have been for some time one of my sorest troubles."

Darwin's concerns were real, and it was a revolution in physics in the early twentieth century that would provide the solution. Ironically, it was a combination of the two problems that Lord Kelvin himself had identified that provided the solution. Radioactivity explained how the Earth could remain hot for so long, and nuclear fusion explained how the Sun could shine for billions of years.

Can similar clues be hidden in today's biology? Or have we finally found all the physics we will need in biology forever? This was what I wanted to know. Dawkins hesitated, unsure, but others are in no doubt. They claim that biology offers nothing of importance to a physicist who is interested in the fundamental laws that govern the universe. Similarly, a biologist can settle for significantly less physics than science already knows.

I think Dawkins had good reason to hesitate.

## The Code

What is life? In 1944, toward the end of World War II, the physicist Erwin Schrödinger (1887–1961) wrote a book with just that title. He approached the issue as a physicist, without ruling out the need for something completely new, even when it comes to fundamental physics, but his book has become historic thanks to a specific idea. He imagined how genetic information could be stored physically, perhaps in the form of some kind of crystals hidden in the cells.

Although at that time he had no access to all the correct details, his theory was largely confirmed when the double helix structure of DNA was discovered a decade later. At a pub in Cambridge on February 28, 1953, the British scientist Francis Crick (1916–2004) told his friends that he and James Watson had revealed the secret of life. Their discovery was entirely in line with Schrödinger's speculation, but it took a physicist to get the biologists on the right track.

The Russian astrophysicist George Gamow (1904–1968) is known for his contributions to cosmology and his acclaimed books on popular science. In a letter to Crick and Watson in the summer of 1953, he suggested that DNA should contain a code in which each word or codon contained three letters specifying how amino acids should be assembled into proteins. Proteins are among the most important components of all living organisms. Gamow's major contribution was to reduce the problem to a mathematical exercise to match sequences of nucleic acids in DNA to the order of amino acids in proteins. Instead of the 26 letters of the English alphabet, the genetic code needs only the four letters A, C, T, and G, which correspond to the nucleic acids adenine, cytosine, thymine, and guanine. Unlike common languages, where there is no specific limit on the number of letters within a word, each codon in the genetic code always contains exactly three letters, such as AAA, AAC, AAT, etc. In this way, you can print 4 x 4 x 4 = 64 different codons of which 61 represent amino acids, while the remaining three signal when to stop protein synthesis. For instance, AAT stands for leucine and CGA for alanine. The order of the codons speaks of the order of the amino acids in the protein that is being produced by a cell. There is a certain redundancy in the language because the number of amino acids contained in living organisms is not greater than twenty.

Why these numbers? Four letters, codons limited to three letters, and codons with only twenty different meanings? It is striking that the code is so universal, from the simplest bacteria to us humans. It shows a unity of all life and points toward a common origin. One could have imagined different forms of life that each used DNA, differently coded, but this is not what we find. Through the billions of years, evolution has changed what is told but not the language itself.

To illustrate the remarkable role that genes play in the living world, Richard Dawkins invented a striking metaphor in the form of the selfish gene. What he wanted to show was how genes form the very heart of evolution. The genes develop and change as a consequence of how the environment affects the organism that the genes give rise to. Genes that are expressed in the form of organisms that survive and multiply are those that are passed on to new generations. Genes that fail to produce offspring that reproduce disappear. The organism in the form of an animal or a plant is secondary and nothing more than a tool for the selfish and ingenious gene.

Does this mean that the genes are in some sense more real than the organisms? One might conclude that while genes exist, the organisms they encode are nothing but illusions. This is reminiscent of how the pre-Socratic philosopher Democritus argued that the only things that really exist are atoms and voids, while everything else, like bitterness and sweetness, is just convention. According to this way of looking at life, genes play the same role in biology as the atoms in physics.

Dawkins's metaphor is compelling, but it's still not really clear what a gene is. Just as an organism is only a way for a gene to express itself, perhaps the same can be said about its physical manifestation in the form of a DNA molecule. When we search for the core itself, we find that genes are nothing but intangible information. Following from this, Dawkins introduced the concept of memes, in which natural selection has an unexpected application. A meme can be a single word, a concept, or an idea capable of spreading from one person to another. It can be a simple phrase that is used by a large number of people, or a system of ideas, even a religion, that affects people in a profound way and determines how they build their societies. Memes can be spread via the internet to successfully infect human brains around the world. They can be passed on to others through the spoken word. The memes can rest in old books for hundreds of years and then be rediscovered and continue their journey around the world. They need their physical form but are completely independent in their content. Pure information can be encoded in many different ways in a human brain, just as ideas may be recorded in ink on a piece of paper or data stored in a USB flash drive.

The discovery of evolution and the genetic code suggests that the very essence of life is pure information, sequences of letters that describe how to put together an organism in the form of an animal or a plant whose sole purpose is to make more copies of itself. But can this really be the case? There is an important catch in the reasoning: A code is meaningless if there is no one who can read it. The human genome, with its sequence of billions of letters consisting of A, C, G, and T, is an example of just that.

It will not be long before many of us will have access to our complete genetic codes. It's not a lot of information. You can easily fit your own sequence on a USB drive, with space left for several of your friends to do the same. Let us imagine how the sequence is sent into space with the hope that an advanced civilization in the distant future will be able to figure out what it is all about. Could they make a copy of you? Not a chance.

It is not just a practical difficulty but impossible in principle. Even if aliens had reason to suspect that it was a code based on the DNA molecule, they would not succeed. Similarly, one might wonder what we could do with a code received from the stars if we had reason to believe that it represented a genetic code. Even though we suspected that it had to do with DNA, that would not be enough to know how to translate the information into a living organism. In short, you may be able to do without a father but not without a mother.

The necessary code key is housed by the complete cellular system that reads and interprets the code and realizes it as a physically living organism. Without cells that can read the code, the DNA molecule remains meaningless. Out of context, it is not much more selfish than any other molecule. In the same way, a piece of computer code is useless if you do not have access to the right kind of computer to compile and execute all commands. With the help of a program developed for an Apple computer, you will not get far if you try to run the program on a PC. What's worse, without the right cables, you cannot even charge the battery of your computer. It takes greater effort to build a physical computer than to write a piece of code. For example, we have long known how to write quantum computation programs even though we still have not succeeded in building a real quantum computer.

The relationship between the genetic code and the processes that read it is like that between the chicken and the egg. Which came first? It is like an encryption where the code required to solve the encryption itself is encoded by the encryption—which in no way makes it easier to solve the incomprehensible encryption. But it's actually worse than that. Not only is the information pointless if you do not know how to read it but there are many indications that DNA does not actually contain all the information. The cell must also know when it is time to use the information in a piece of DNA to build the proteins it encodes. This information is not necessarily stored in the sequence itself. For example, it has been found that DNA can be chemically altered without altering the sequence of letters, so that it is not expressed. This leads to the possibility that acquired properties of living organisms are passed on through the generations and therefore goes against the message of what is usually called the central dogma, according to which information can be transferred from DNA to the proteins but no information can be transferred back to DNA.

There are more ways for an organism to take advantage of its experiences and pass them on to future generations. We humans have invented a particularly effective method that has been in use for thousands of years. Information about what we have experienced and learned during our lives is stored in our brains in the form of memories that can be transferred to our children's brains through education. Nowadays, it is common for the collective experience to take the detour via books and electronic media before it ends up in the recipient's brain. Like genes, memes play an important role in how humans evolve and adapt. The time scale differs, but from a fundamental physical standpoint it is the same thing. It is a scientific problem to find out how the information is stored and changed, and there is no reason to believe that we have all the pieces of the puzzle.

Information is a useful concept when describing how living organisms function and develop, but it is a mistake to think that in this way one captures life as a physical phenomenon. It's much more interesting than that.

## Living Machines

We, as humans and scientists, love simplicity, and are happy when we find it. For a physicist, beauty is nothing but unexpected simplicity. We imagine the universe as a brilliant machine, constructed according to mathematical principles where essentially different phenomena are governed by the same basic laws. Previous generations linked this to the presence of a divine engineer. What characterizes a machine is its design, which shows that it was constructed and assembled by an inventor with a specific intention. There is a meaning to the machine that can be traced back to an external agent.

As a scientist, it is natural to suffer from a weakness for reductionism. After all, science is to a large extent the search for patterns that suggest a hidden simplicity. But in modern biology, there is no need for beauty or perfection in a world that lacks a creator. "Good enough" is life's slogan, and sometimes it's just barely good enough. There is no intrinsic value in a living world constructed in a simple and pedagogical way. The organisms of biology therefore differ in a fundamental way from machines. They arise through evolution, where there is no design, and the watchmaker is blind.

The difference between organism and machine already worried the philosopher Immanuel Kant (1724–1804). In his *Critique of Judgment*, he argued that there can be no purely mechanistic explanation of what life is. One might wonder how he could know that. His insights into biology were presumably quite limited, given that his book was written long before Darwin discovered evolution. What Kant sensed was a tension between the mechanistic and teleological explanations. An organism is both cause and effect in itself. The parts create the whole and the whole determines the parts. Kant was pessimistic about the extent to which this was something we could ever understand. He worried that "the organization of nature is . . . not analogous to any causality known to us."

In light of this, a risky contradiction emerges in the way we tell the history of life. When you insist on machine-based images, as many biologists do, you inadvertently lead your thoughts in the direction of intelligent design. The metaphor of the selfish gene collapses when it is pushed too far. There is a deep irony in the fact that the image we paint, with a code that is read and implemented, is so similar to the confused ideas about intelligent design that we may want to oppose. The living world does not work that way. There is no clear boundary between the code and that which interprets the code. The genome does not consist of intangible information. It consists of matter and is part of a cellular system that has evolved over billions of years without a need to fit into simplified models.

This is not something that worries the American philosopher Daniel Dennett. In his book *From Bacteria to Bach and Back* (2017), he does not shy away from speaking in terms of constructions in nature as long as it is emphasized that there is no constructor. He would not find interesting the distinction between machine and organism that I try to make. He finds it unprofitable to spend time convincing laypeople that there is no design in living organisms. One should instead fully accept it as an obvious fact that there is design everywhere and explain how it arose through blind evolution. The deputy designer is natural selection in combination with new gene variants arising through random mutations. The organisms that succeed in surviving and multiplying pass their genes on to future generations and in this way the living world evolves. I have some sympathy for his attitude. I completely agree with how he describes evolution and I completely agree that nothing beyond physics is needed to describe either life or consciousness. Still, we differ on one crucial point: What do we really mean by physics?

Dennett quotes colleagues who express themselves more cautiously than he does. He complains about how their language is weighed down by pointless excuses. In the end, he may be right, but if we take modern physics as we know it very seriously, his position is not at all consistent with our understanding of the fundamental laws. There is simply no place for meaning or teleology—not even in the form of analogies.

Physicists like myself sometimes believe that we carry the whole of science on our shoulders. Biologists may be able to get away with the fact that they are only dealing with models that are not really to be taken literally. The difficulties can always be shoveled downward, until they end up with the physicist, who has no one else to blame and is left empty-handed, with equations that can offer neither meaning nor teleology. As long as we focus on the outside world, we are free to use a language loaded with meaning. Like Dennett, I can see the pedagogical points.

It is through stories that engage emotionally that the world can be made comprehensible. But here we want to achieve more. Our ambition is to include the subjective inner world as part of the physical world. If everything is physics, this is inevitable and there is nowhere to hide. The problem is that physics usually leaves the physicist out of the equation. There is a crucial line that must always be drawn, which separates the observer from the outside world. The whole construction collapses when we accept that we as observers, whether it is you, me, or Daniel Dennett, are bodies of flesh and blood that follow the same laws of nature that we so desperately try to understand.

In his book, Dennett recalls a lecture in front of a large audience of physicists. He asks them how many people understand the meaning of $E = mc^2$. Everyone claims that they do, but a theoretical physicist stands up and argues that this is not true at all. Only the theoretical physicist really understands what it is all about. Dennett's point with this story is that understanding is a relative concept with many shades of gray. My point is another. I am one of those who actually understand Albert Einstein's (1879–1955) formula and what makes it so remarkable. Along with this understanding comes an insight into what physics really is, what it can accomplish in its current form, and what its limitations are. There is an honest clarity in physics and a precision in the terms used that help to expose its limitations. This clarity and precision is lacking in biology. There is a difference between created machines and evolved organisms that is not fully defined. Evolution shows how life is pragmatic and focused on results, in stark contrast to physics, which still celebrates simplicity and beauty. There are empirical reasons to believe that there is something special about life that our models have not yet managed to capture.

This also applies to the subjective inner self, which can be assumed to be represented in other forms of life, as well.

I think back to the muddy, smelly ponds from my childhood where I collected strange creatures in the years before I began to study the stars in hope of finding more promising ways to explore the deepest mysteries of the universe. Could it be that it is among the dragonfly larvae in the mud that the biggest riddles are hidden?

# 3

# The Universe Is Not Mathematics

*We have at any rate one advantage over Time
and Space. We think them whereas it is extremely
doubtful whether they think us!*

—JOHN COWPER POWYS

PHYSICISTS, IN PARTICULAR THEORETICAL physicists
like myself, often suffer from a Platonic complex. We
are constantly amazed at how effectively mathematics can
be used to describe the world and we're happy to tell everyone we meet. Theoretical constructions based on mathematical reasoning, where simplicity and beauty are used
as guiding principles, turn out not only to reproduce what
one already knows but to give rise to completely new, and
surprising, predictions. Theoretical research is often about
using the mathematical world of ideas to understand the
world itself. The first real great triumph came when Newton mathematically succeeded in connecting the physics of
celestial bodies in motion with how apples fall from trees. It
was not just an insight that satisfied his curiosity but also a
practically useful theory that laid the foundation for technological development hundreds of years into the future.
The theories of relativity and quantum mechanics use even
more advanced mathematics to successfully describe what

we observe in nature and predict the outcome of experiments. Everything we find in the interior of matter in the form of particle physics can be successfully incorporated into a mathematics that, however high-flying and advanced it may seem, is nevertheless characterized by an astonishing simplicity. When we are faced with this, the temptation is to see mathematics as something existing independently. How else could it work so well?

In the philosophical tradition, this worldview is called Platonism. And for those of a religious persuasion, it is not difficult to grasp. The very concept of natural law already contains a clue to how one should think. Nature is forced to follow laws established by someone or something that is above and beyond nature. In other words, God. This God is obsessed with mathematics and has dictated how everything is connected. Revealing God's will was also the driving force of many physicists in previous centuries. Replacing God with Nature, with a capital *N*, does not really change much.

For some of my colleagues, even if they would never admit it, mathematics still fills this almost religious role—not necessarily in the form of any explicitly divine tool but entirely in itself. Many are likely to find comfort in its supposedly independent existence, which is reminiscent of how a religious person views God and the Kingdom of Heaven. One can liken the belief or nonbelief in a God to the belief or nonbelief in a transcendent Platonic idea of mathematics. In none of the cases is there any good empirical evidence for what is claimed. Depending on the different roles that mathematics and beauty play in various fields—although I cannot base my claims on any scientific research—I suspect that there are more religious believers among mathematicians and physicists (especially astronomers) than among biologists.

## The Multiverse

Max Tegmark is a Swedish physicist and cosmologist active in the United States who has roots in the region of Dalarna in central Sweden, just like myself. I have had several interesting and entertaining discussions with him over the years and it is clear that in many ways we share each other's views of the universe. We both suspect that our world is much larger and more diverse than our instruments have so far managed to explore.

The small part of the universe that we have some control over extends all the way from the room where I sit and write this book past the stars in the Milky Way and past every galaxy we can observe, all the way to the cosmic horizon. How far is it? It depends. If we look back as far in time as it is at all possible, we can discern in the cosmic background radiation structures as they appeared 13.8 billion years ago. When the light from these structures began their journey, they were just a little more than 40 million light-years away, but in the billions of years that have passed, they have evolved into galaxies that, through the expansion of the universe, are now almost 50 billion light-years distant. We cannot see what these galaxies look like right now, but we have good reason to believe that they are not too different from those close to us.

Although this is already large, there is no reason to assume that the universe does not extend even farther. It is likely that there are galaxies and stars scattered over much greater distances. Of course, there may be rare phenomena that have never occurred in our corner of the universe, but the rules of the game can be expected to be the same—including the fundamental laws at the level of particle physics. We do not know how far we can travel before something fundamentally new happens and we reach some kind of limit. It's slightly reminiscent of computer games like No Man's Sky, where you can continue your journey of discovery in a seemingly endless world.

But if you travel really far, it is not at all an unreasonable guess that the rules of the game themselves will eventually change, and that what we mistakenly view as universal laws of nature differ in a radical way. Of course, it would be difficult to physically travel into these alien worlds. If the laws of nature are different, the atoms in your body would simply dissolve.

Throughout history, man has repeatedly discovered that the world is much larger and more diverse than previously thought. Seafarers found new seas and new continents, until eventually the whole Earth was mapped. We now know that there are countless other planets just in our own galaxy. What the universe looks like, and how far beyond the cosmic horizon the universe extends, we do not yet know.

Making a distinction between the part of the world that we think we have a decent grasp of and what we can so far only guess at is how the concept of multiverse has become popular. The idea is that our universe is a small part of something much larger, a multiverse, which remains for us to explore. The important thing is not whether the different parts of this multiverse are completely separated, and in that sense correspond to different universes, or whether they are interconnected. The important thing is that there is much more out there than we have knowledge of yet. This way of thinking follows an old and fairly successful tradition of speculative geography. The goal is simply to find out how far the physical world extends.

Multiverses can also shed light on why our universe looks the way it does. It is strange, after all, that the laws of nature are so finely tuned that there can be stars, planets, life, and thinking beings like us at all. An alternative to the notion that everything is the way it is due to some form of intention is that the universe, or rather multiverse, is so large that as a consequence of pure chance the conditions should just happen to be the right ones at least somewhere. The reasoning is similar to how we view the fact that the Earth is such a hospitable planet for life. This is not at all about a higher power deliberately adjusting the Earth's orbit and composition, but about the fact that there are so many planets in the universe. We simply exist on a planet where enough happens to be just right.

Some support for these ideas comes from string theory, which is an attempt to unite the different parts of known physics and resolve the troublesome contradictions that exist between the general theory of relativity and quantum mechanics. In these attempts all sorts of mathematics are used and it seems more or less inevitable that there must be more dimensions of space than those we can observe. From a strictly theoretical point of view, it is also reasonable to imagine natural laws other than ours and perhaps also a multiverse that connects them. So far, we do not really understand the mathematics, and there are no useful predictions. As earlier in history, it is likely that new observations and ingenious experiments will determine when the really big advances and discoveries come. Maybe it will also turn out that we lack crucial pieces of the puzzle.

Max and I agree that the existence of a multiverse of this kind is not unreasonable. The problem is that Max, and he is hardly alone, is not content and believes that beyond all this there are worlds of an even more wonderful kind.

## The Parallel Worlds of Quantum Mechanics

Among the strangest things you can come across in physics is quantum mechanics. It shows that the way the universe is at a given moment in time is not unique, but in a superposition of different possibilities. It's not something we usually need to worry about so much, but for small particles it's part of daily life. An electron in orbit around an atomic nucleus cannot be said to be in a certain specific position moving in a particular way, but must instead be described as a smeared wave. This is absolutely crucial for understanding the structure of atoms, and all the experiments that have ever been carried out confirm that this is exactly how the world works.

Matter thus has a kind of double nature, where it is sometimes best described as waves and sometimes best as particles. How the particles and the waves relate to one another is something that is sorted out in quantum mechanics. A little carelessly, you can say that matter behaves as waves when you do not look at it and as particles when you do look. The wave describes the possibilities that exist, of which only one is realized once you perform a measurement. All the other what-ifs disappear and only what really happened remains. The strength of the wave gives the probability of one or the other result.

It is important to realize that no choice is made until a measurement has been performed. The wave nature is essential and leads to a number of phenomena that can be observed and measured when dealing with individual particles. It only becomes troublesome when you apply quantum mechanics to large objects. Everything big consists of small things and it would be reasonable to assume that the laws of quantum mechanics also apply to large objects. For example, cats. In a famous thought experiment, Erwin Schrödinger imagined how a cat is in a state where it is dead and alive at the same time. The reason for this is that the cat's fate is linked to whether a certain atomic nucleus has decayed or not. As long as nothing happens to the atomic nucleus, the cat lives on, but if it decays, the cat is exposed to a deadly poison. Since the atomic nucleus can exist in a state of a little bit of both—decayed or not decayed—until it is observed, the same must apply to the cat. All this happens inside a box and it is only when we open the box and look that the case is determined. Is the cat dead or alive? Quantum mechanics suggests that it is both—until you open the box. Can something so absurd have anything to do with reality?

For a long time, the mystery was this: When is the choice made between the possible outcomes of an experiment? Or, in more mathematical terms, when does the wave collapse and gather around the final result, which is actually measured? There have been many suggestions for a solution. Some, such as the Danish physicist Niels Bohr (1885–1962), have argued that one probably does not need to worry about how it really works. There is a difference between big and small things. The collapse occurs when the system you want to study is connected to a large and clumsy measuring device where quantum mechanical peculiarities are irrelevant. If you just follow the quantum mechanical recipe, you can be sure to obtain predictions that correspond to reality. What more could you ask for? "Shut up and calculate" is the motto you usually refer to. This way of interpreting quantum mechanics is still quite popular and is often called "the Copenhagen interpretation"—it was in Copenhagen that Bohr worked. However, there is a problem of principle, which in time has become a practical problem as experimental techniques have developed. Where, exactly, should one draw the line between the small and the big?

There are other physicists and philosophers who have taken a more dubious view, linking the collapse to the presence of the consciousness of the observer that intervenes and forces matter to choose. There are still others who have been looking for new physical processes that will give rise to a collapse—perhaps something to do with quantum gravity. Now most physicists probably agree that none of this is needed. The collapse occurs automatically as soon as the object is large enough, and this takes place for a very simple reason. A system with a few electrons can easily be isolated from its surroundings, while this is not at all so easy when it comes to large objects such as cats. They are in constant interaction with the rest of the universe, which in practice performs the observations required no matter what we do. In this way, there is a natural transition from the microscopic world of a little bit of both to a more well-defined macroscopic world.

In a fundamental way, modern physics—including quantum mechanics and relativity—do not differ from the old classical physics. Given an initial state of matter, the laws of nature can be used to deterministically calculate and predict the state of matter at a later time. This is obvious in Newtonian mechanics and true also in the quantum world. A quantum state, sometimes described with the help of a wave function, develops over time with the help of the deterministic Schrödinger equation. It is only when we perform a measurement that we need to introduce probabilities and determinism partially fails. The reason is purely practical. To be able to talk about measurements, we must separate the object from the rest of the world and put it in focus. The connection to the surrounding universe causes information about the system to leak and be lost. In this way, chance and probability creep in. If we were to abstain from measuring anything at all, our quantum mechanical description of the universe would be completely deterministic. The price we would have to pay is that nothing would actually happen within the framework of our model.

The result is a very satisfactory picture of what happens in a measurement and there should not be much more to add. But there are many who simply cannot accept this. Quantum mechanics has confused physicists, and perhaps philosophers even more so, for decades, and it seems difficult to let go of the idea that there could be something new and unknown lurking behind the scenes. If you insist on wanting to avoid the problems associated with the collapse, one way is to simply deny that it takes place at all. Measuring and collapsing the wave function would thus be an illusion that follows from a limited perspective. In reality, no choices are made. That was what the physicist Hugh Everett (1930–1982) proposed in the late 1950s in the form of *parallel worlds* (sometimes called *many worlds*), where no choices are made but everything happens. At every moment in time, history splits into different versions that represent all the possibilities allowed by quantum mechanics. The world you are a part of is not special. There are myriad other worlds where copies of you consider their particular variant to be the real one. No one is right, no one is wrong, and every world is equally real. Note that this has nothing to do with the kind of multiverse we have previously come in contact with. The multiverse is just the very reasonable, and rather conservative, assumption that the only truly existing world is much larger and more diverse than we have hitherto had reason to suppose. The hypothesis of parallel worlds, on the other hand, states that our own as well as the entire history of the multiverse branches off at every moment so that everything that can happen actually does happen.

Strange? Maybe, but the important question is whether the model gives a correct picture of reality. Many would agree and claim that it is even a better interpretation than the one coming from Copenhagen. It can at least be a practical way to organize your thoughts and calculations. The question is, however, is that how it is? In reality?

Those who profess a belief in parallel worlds have no problem with how the surrounding universe keeps measuring and carves out the various possible stories. This is exactly what forces the cat to be either dead or alive but not both at the same time in the same world. The extra strange twist that is added is the claim that no choice ever takes place and that all the alternatives continue their own existence in the form of the various parallel worlds. In one world the cat is only dead, and in another world that is just as real, it is only alive.

We theoretical physicists have the privilege of claiming all sorts of absurdities, and getting away with it, by giving the impression that we are the only ones who understand how it really works. Sometimes that may be true, but when it comes to parallel worlds, it is just as crazy as it seems.

What is central to the theory of parallel worlds is the question of what is considered to be real. Are the mathematical tools part of a model that we use to make predictions, or do they correspond to something independently existing? How one approaches that question is decisive for what conclusions one draws. If you believe in mathematical structures as something real that exists independently of us, the step is short to put the wave function in this category and confess to parallel worlds. If you, like me, believe in the existence of a material universe independent of mathematical formalism, there is no reason to believe that there is more than one reality.

Let's return to Max Tegmark. For him and other like-minded people, parallel worlds are real. He weaves them together with the geographical multiverse and the Platonic world of mathematical structures into a single unit. Is it beautiful? In any case, it is a popular idea that appeals to many and resonates in popular culture. Through parallel worlds, physics, in the eyes of many, gets a chance to improve its reputation. Physics claims that there is no afterlife, no meaning, only the dead atoms of Democritus, and it has been accused of disenchanting the world. Through the theory of parallel worlds, physics gives us an escape route out of this single reality that feels too small and cramped, and thus inspires hope. What if all your dreams could come true in one of the parallel worlds? "Believe in them and you will be saved!" is the joyful message.

Personally, I'm terrified. I understand Johannes Kepler (1571–1630), who got scared when he heard how Giordano Bruno (1548–1600) speculated about a large number of worlds and about how each star was a Sun. Admittedly, fear is not a reason why there should be no other worlds than ours—there are many horrible and real phenomena in the natural world that can lead to nightmares. Not the least of which are black holes. But my fear is of a different kind.

I receive questions about physics via email almost daily and I am far from able to answer them all. This is a constant source of bad conscience. But a few years ago, I received an unusual letter that gave rise to an exchange. The sender was a woman who had read Max's book and been frightened out of her wits by the concept of parallel worlds. It was a very disturbing letter. She took the playful thought experiments that Max and I had fun with in all seriousness.

For her, it was literally a matter of life and death. She was particularly concerned about a thought experiment on quantum mechanical suicide. Suppose you play Russian roulette with a revolver pointing toward your head. You have loaded only one shot into the revolver, but there is a great risk that one shot will fire and that you will die when you squeeze the trigger. You spin the cylinder and try again. You may be lucky, but sooner or later it will go off. What would you experience if the theory of parallel worlds were really true? In every moment you are split into different copies and there will always be successful versions of yourself that survive no matter how many times you try. One of these will be you—otherwise, you would not be there at all. The reasoning follows Epicurus's thesis that where death is, we do not exist, and where we are, there is no death. So why not take the chance and give it a try? Admittedly, there would be worlds, an infinite number moreover, where desperate relatives would find your dead body, but it is not in any of these worlds you would end up. And really, there is no reason why you should have to play Russian roulette— we play the same kind of game just by living our lives. The theory of parallel worlds predicts that from your own perspective you will with absolute certainty live as long as it is physically possible. If there is a probability that you will live for two hundred years, for a thousand years, or for a million years, you will do it. If there is the slightest probability that you will never die, you will have eternal life.

The woman who wrote the letter demanded to know whether this could be true. To give her peace of mind, she wanted convincing arguments as to why there can be no parallel worlds. The choice between an infinite number of worlds and a single one was not something she could take lightly. My response to the desperate writer of the letter consisted of two parts. The first piece of advice was not to take people like myself and Max too seriously. The quarrel between physicists about phenomena that go far beyond what we can measure only shows that we have no idea. Philosophers who try to interfere in issues that ultimately have to do with physics are often not very helpful, either. Since I did not feel completely satisfied with writing so derogatorily about my own authority, I added some hand-waving arguments as to why I was right and told her that I was writing a book on the subject. I hoped she could patiently wait for it.

What would happen to the world if we were all convinced that the parallel worlds of quantum mechanics are real? I'm sure it would be really bad and that the woman who wrote me the letter was on the right track. Individuals would lose their foothold and the result would be a collapse of our civilization. Ironically, those who believe in parallel worlds could still find solace in the idea that there must be worlds where one does not allow oneself to be convinced that the theory is correct and therefore can continue to live one's life as usual.

Physics is ultimately about reality, and there are answers that are right and answers that are wrong. It may be that it will never be possible to find out if there is one world or many, and philosophers can forever continue to discuss arguments for or against the different positions. For my part, I believe that the difference is absolutely crucial for us humans as meaning-making beings of flesh and blood.

Phenomenologists such as Edmund Husserl (1859–1938) long ago pointed out a possible solution to the dilemma. It is not enough to realize that dualism is unsustainable, and that all our thoughts and models—including science itself—are embodied. The world itself, of which we are a part, is not only represented in our consciousness; it also presents itself. What we directly experience with our physical senses through our organic bodies is the world itself in a physical and concrete sense. There is a difference between models, simulations, memories and what is real. According to the French philosopher Maurice Merleau-Ponty (1908–1961), reality has an essence that differs from what we only imagine. Long before any mathematical modeling of the world lies a basic scientific observation that lays the foundation for everything else: The world exists.

Let me now try to convince you that Platonism is not a viable path—at least not if you want to base your worldview on a scientific approach.

## Is It All Math?

According to the mildest form of Platonism, matter is governed by mathematical laws that lie outside of the material universe, in the world of ideas. Such a worldview is dualistic in nature and presupposes a perspective where one as an observer can stand outside the world and control it. This is entirely consistent with, and almost assumes, a religious worldview in which God, as the great mathematician, instituted the laws. Man, or at least the mathematically trained physicist, can, like God, stand outside the world and try to read the thoughts of the Almighty. Belief in the Platonic form of mathematics can thus be likened to a belief in God. Rational arguments against such beliefs are not always effective. If you like mathematics but do not really feel comfortable in the dualistic camp, you can, like Max, try to throw away matter and completely rely on mathematics. This leads to an even more serious form of Platonism.

The idea that everything is mathematics can seem seductive when we think about phenomena in particle physics and the quantum world that are far from what we usually deal with in our daily lives. The language we use to describe how particles and fields behave uses sophisticated mathematics that in many ways contradicts our everyday intuition. Since the concepts we deal with in physics—for example, quarks—are completely determined by numbers, and the quarks correspond to something truly existing, it follows from this reasoning that reality is nothing but numbers. But if we think of phenomena that are a little easier to grasp, extreme Platonism immediately ends up in difficulty. Let's take a planetary orbit as an example. In what way is the orbit really an ellipse? The point is that the planetary orbit would not only look like an ellipse, when you compare it with your notion of an ellipse in your mind; it would really be an ellipse in a concrete, material sense. Does this mean that the numbers that describe the planet's speed and position are out there in space? For real? In which units are they expressed and in which coordinate system?

Electromagnetism can serve as another illustrative example. A complex number consists of a real part and an imaginary part. The imaginary part is proportional to the square root of -1 and is denoted by $i$. Every real number, regardless of whether it is positive or negative, has a square that is positive (or zero, which is the square of the real number zero). However, we have by definition that $i^2 = -1$. Complex numbers play an important role in the case of alternating current where the phase of the current, or voltage, is described by a rotating arrow in the complex plane of numbers. It is not particularly advanced mathematics or physics and is an integral part of the technology around us. Does this mean that the square root of -1 appears when you turn on a light switch? If you instead choose to describe what happens without this imaginary $i$ (which works just as well), does that mean that this type of mathematics is suddenly not present? If one sees mathematics as a tool to describe the world, none of this is particularly strange. If you want to see the world itself as pure mathematics, it is paradoxical.

What is claimed to be the best argument for the existence of mathematical objects was put forward by the great American philosophers Willard Quine (1908–2000) and Hilary Putnam (1926–2016). Their reasoning is based on two statements that they believed led to the conclusion that mathematical objects exist.

First, everything that is indispensable to a scientific theory is also real.

Second, mathematics is indispensable to scientific theories.

Thus it follows that mathematical objects must exist. This is the whole argument in a nutshell. It is important that the word *exist* is taken very seriously. It is used in a sense that refers to something that has an existence independent of us as observers. Following Quine and Putnam, one concludes that mathematical objects existed before humans walked the Earth and perhaps even before our universe was born. It is simply old-fashioned Platonism.

Is this an inevitable conclusion? Both claims on which Quine and Putnam based their argument have been questioned. Perhaps the most drastic possibility is to question the second statement and claim that it is possible to engage in science without mathematics—as the American philosopher Hartry Field did in 1980 when he argued that mathematics is pure fiction, albeit useful. Does this make sense?

Let's say you are watching a game of soccer. A player is about to shoot the ball toward the goal and aims at the top right corner. How hard should she hit the ball and in what direction? What is the optimal speed? As a physicist, I would approach the problem by setting up Newton's laws and using my mathematical ability. I would need to take into account the mass of the ball and how the mass is distributed. How the ball rotates and how it is affected by the air. Isn't it a bit windy, by the way? Then that's another complication, one among many, that I have to consider. The problem is quite complex if one is to perform a realistic calculation, and it is something that I would never dream of giving as an exercise in a university course in basic mechanics.

Long before we even get started, long before we write down a single equation, the player has already made her decision and the ball is on its way. If she is a skilled player, it is likely that the ball will reach its target. However, no one watching the game will be particularly interested in what our calculations, which may last for several more hours, will lead to.

Is mathematics really indispensable for analyzing problems such as the trajectory of a ball? It does not seem very likely, given that you hardly need a university degree in mathematics or physics to be good at soccer. But perhaps the fact that mathematics does not seem to be needed is just an illusion. And the player unconsciously goes through a mathematical calculation in biological circuits that evolved over the ages, fine-tuned after hours of training. Could it be that not only our soccer player but every living creature that has to deal with moving objects, such as a swallow catching a fly, uses methods similar to those invented by Newton in the seventeenth century?

You might try to argue that mathematics is indispensable for the physicist but not for the soccer player or the swallow. But if there are alternative methods, how can one of them be considered indispensable but not the other? To get around this, one could try to argue that there is a difference between playing soccer and practicing science. A popular and reasonable method of assessing science is to examine how effective and successful a model is compared to reality. In the case with a ball that is shot over a soccer field, it seems that a skilled soccer player can handle a long series of fairly advanced, and successful, computations.

One can, of course, argue that this is a special and very limited example. The claim that mathematics is indispensable refers to much more general contexts. Surely it is impossible to do without mathematics if you want to map the motion of planets or what happens near a black hole. But is it really true? There is nothing that basically forbids the existence of extraterrestrial beings who feel just as at home among planets and black holes as the soccer player with the ball. In the TV series *Star Trek: Discovery,* the starship sometimes encounters space whales, or gormaganders, which eat alpha particles in stellar winds. What kind of abilities to handle exotic physics could such creatures have developed? As far as the question of the indispensability of mathematics is concerned, it does not matter whether such beings really exist or not. The very idea of such a possibility casts doubt on the correctness of the assumption.

One can also ask how the laws of physics determine the movement of the ball on its way to the goal in practice. Does Nature, or whoever it may be, perform lightning-fast calculations to make sure the ball chooses the right course? Is mathematics indispensable to Nature? A much more reasonable assumption would be that there is no mathematics—not in the true sense of the word—when the planet moves around the star, or the swallow catches the fly, or the player scores. Mathematics arises only in the brain of a poor physicist who tries to understand what is going on.

Let us now, just for the sake of argument, follow Quine and Putnam and assume that mathematics is necessary. Does this mean that mathematics exists out there in the real world? Not if you ask those who actually use it. One can, for example, consider a more specific mathematical concept as infinity. Does it exist? In physics, infinity is often used to make calculations easier. For example, if you want to describe waves on a water surface where the depth is great, it is practical to assume that the water depth is infinite. It makes no difference anyway. In reality, of course, the depth is not infinite; it is just easier to pretend in the mathematical calculations that it is. Physicists are lazy and use the mathematical model that does the job with as little effort as possible. "Consider a spherical cow" is the classic phrase, used to make fun of attempts where you try to simplify a bit too much.

It is thus far from clear that one can argue that mathematical concepts exist based on their usefulness. What happens when old concepts are replaced with fundamentally new ones? Do the old ones suddenly cease to exist and the mathematical world is redrawn? The point was, after all, that mathematical concepts would exist no matter what we humans invent.

What applies to mathematics also applies to what we call natural laws.

## There Are No Laws of Nature

The first thing you learn about physics is that there are laws of nature. Matter may not behave as it pleases, but is under constant surveillance and forced to follow certain established principles and laws. There are many metaphors and parables to be found throughout the history of science that are connected to this. Even the not very religious Albert Einstein escaped with "God does not play dice" when he was to demonstrate his hesitation before quantum mechanics.

But the laws of nature, like mathematics, belong to

our description of the world and are not at all something that needs to have any existence independent of us. This has nothing to do with any kind of subjectively oriented relativism. I am completely convinced of the objective existence of the physical world and find it a meaningful occupation—I even make a living from it—to construct models that, as far as possible, mimic how the universe behaves. It is in such models that the laws of nature have a function. The universe is not governed by what we call the laws of nature; rather, it is the laws of nature that are constructed by us to follow the universe.

We can take a falling apple as a simple example. In Newtonian mechanics, one cannot do without the concept of force. It is difficult, almost impossible, to understand what is happening without forces. This is especially true of gravity. The Earth's gravitational force pulls on the apple and causes it to fall faster and faster to the ground. I would suspect that Quine and Putnam in this perspective would have had to admit that the force of gravity thus actually exists. And even today, engineers who build bridges and spaceships use the same equations for one simple reason: They work. But this does not mean that these forces and all the associated mathematics exist independently of us in space.

Ironically, in the general theory of relativity, it is more or less an axiom, an unshakable starting point, that gravitational forces do not exist. An apple that falls to the ground is not affected by any gravitational force, but follows the curved space-time in accordance with general relativity. That is precisely the finesse of general relativity. The apple falls because it sluggishly follows the curvature of space-time, not because a force affects it. There is simply no force.

Newton says that there is a gravitational force while Einstein says that there is none. With Einstein's theory the success of Newtonian mechanics is reproduced, while the accuracy of the predictions is increased. This is, among other things, necessary if a GPS is to give the correct position. Future physicists may find even more effective descriptions. How can that be? The answer is simple. The laws and mathematics can develop and differ because they are inside our heads. In Newton's, in Einstein's, and in yours. Nature does not need physics or mathematics to calculate how an apple falls. The apple just falls, while our description of what happens develops and improves over time.

The Quine-Putnam argument is considered the strongest for a Platonic world of mathematics, but it does not seem very convincing. There is not much support for the fact that mathematics exists outside our brains and independently of us. A concept such as infinity manifests itself, like other concepts, in our organic brains and is connected to our experience of moving and being in the world. And it's probably really just as well. It has been shown that there are snakes in the mathematical garden.

## Hilbert's Dream

After the turn of the last century, the German mathematician David Hilbert (1862–1943) set out a goal that still seems quite sensible. It was not a particularly easy task, a bit nerdy, but someone had to deal with it. His ambition was to create a solid and stable foundation for mathematics. In order to be able to trust mathematics, he believed that it must have no contradictions and be complete. In our technological society, where we rely on mathematical algorithms running on powerful computers, it is easy to feel sympathy for what Hilbert wanted to achieve. Why would we let our lives hang on systems based on mathematical calculations if mathematics itself is not reliable?

If you can prove that a certain mathematical statement is true, freedom of contradiction means that you cannot at the same time prove that it is false. In mathematics, nothing can be true and false at the same time. If there was a single exception, you could show that anything is both true and false and the whole framework would collapse into a meaningless mess of symbols. Hilbert saw that freedom of contradiction was a necessity but was not satisfied with it. There was the possibility that there could be well-defined mathematical statements where it would not be possible to determine whether these were true or false. According to Hilbert, mathematics must be better than that and be able to determine the truth on its own. Mathematics would thus be complete and not need any input from anything outside it.

Freedom of contradiction and completeness were assumed to be the ultimate dream and desire of all mathematicians. If you transfer this to everyday language or thinking in general, the dream becomes extreme. Let us assume that English, or whatever language you want, would be noncontradictory and complete in Hilbert's sense. This would mean that everything you can formulate in words must be true or false and it would also be possible to determine which by pure thinking power. The Earth moves around the Sun. True or false? Is Bach's music beautiful? Does my neighbor have a cat? The whole point of a natural language is that it refers to something outside itself. The Sun, Bach, and my neighbor's cat (yes, it does exist) are things that exist beyond language, with properties that cannot be derived from the words themselves. Hilbert would certainly not have claimed that English is complete, but why would mathematics be different?

Hilbert saw mathematics as pure formalism. The creative

act of finding a proof, which is what mathematicians in general do, is nothing more than a pure and thoughtless manipulation of symbols. Just as when horses were replaced by tractors, Hilbert tried to make mathematicians obsolete. One would not have to refer to anything outside of formalism and thus there would be no meaning in mathematics. Another way of formulating it would be to say that semantics has no place—it's all about syntax.

## Russell's Paradox

Above all, the one who accepted Hilbert's challenge to formalize mathematics was Bertrand Russell (1872–1970). Together with Alfred North Whitehead (1861–1947), he decided to write the ultimate rule book for mathematics, which would dispel any doubts about the basics of numbers and mathematics. The project was doomed to fail. As early as 1901, several years before their main work, *Principia Mathematica,* was completed, Russell made a terrible discovery, indicating that there could be a fundamental error in mathematics. It was a snake in the garden that threatened Hilbert's dream. What worried Russell may at first glance seem like pointless wordplay, not to be taken seriously, but it would doom the entire project.

Let's follow Russell in his reasoning and play a little with sets. A set is a collection of things of a given kind. It is easy to find examples because almost everything you can think of is some kind of set. For example, the set of all oranges in your kitchen, the set of all apples that have ever existed, the set of all elephants that speak Swedish, or the set of all philosophers named Bertrand. There are all kinds of sets, many of which may be empty.

Sets prove to be extremely useful when defining what an integer is. Russell managed to reduce everything one can say about integers to a small set (*sic*) of axioms. With the help of these, he was able to prove, in a rigorous way, all the theorems of number theory without any loose ends. To Hilbert's delight, everything turns into a purely mechanical procedure where nothing could go wrong. With the help of formalism, one could even formulate a proof of $1 + 1 = 2$ where nothing is taken for granted. I have at some point had reason to study what such a proof amounts to and I must admit that I do not find it particularly illuminating.

Unfortunately, Russell stumbled upon a disturbing paradox about sets that put him in a state of shock. The point is that there is nothing to prevent a set from containing itself. Of course, the set of all apples does not contain itself. The set of all apples is not an apple. When it comes to the set of all sets, it is a different matter. The set of all sets is, of course, a set and, confusingly enough, it must by definition contain itself. I say it again: The set of all sets contains itself. It's a little difficult to create an image of this in your head, but it does not matter. The reasoning is simple enough to convince that such a beast exists. Whatever the word *exists* really means in this context. But what about the set of all the sets that do not contain themselves? Now everything has gone crazy.

Russell formulated the question in the form of a famous thought experiment. We imagine a city where there is only one barber. This single barber has the task of shaving everyone who does not shave himself but no one else—something that at first seems quite natural. The simple question that complicates everything is "Who shaves the barber?" Let us test the two possibilities that exist. We start by assuming that the barber shaves himself. But this is against the city regulations; the barber must not shave himself! Second attempt: The barber does not shave himself. And this is also against the rules—the barber must shave everyone who does not shave himself!

Once you have got the idea, it is easy to construct many other similar examples. Feel free to try. The key is self-reference, where important statements refer back to themselves in a surprising way when you follow them a few steps. For example, what do you think will happen if Pinocchio says "My nose will grow"?

Problems similar to those of the lone barber are of exactly the same kind as the set that does not contain itself. Just replace the word *shave* with *contain*. You might think it's just a silly play on words, but as we've seen, set theory is serious. It is the very root of the project that would lay a solid foundation for the whole of mathematics, and one cannot take the paradox lightly. Russell tried to find a way out but failed, as we shall see, for good reason.

Incidentally, set theory risked overturning my own burgeoning mathematical career in primary school in the early 1970s. According to new modern pedagogy, sets would be used to teach mathematics under the slogan "Hello mathematics!" The tasks were often to draw circles marking one or the other subsets of different kinds of objects of different colors. "Indicate the subset of the balls that are red!" Since I am color-blind, I had a hard time knowing if the ball was red or green and which set it actually belonged to.

## Gödel Shatters the Dream

The genie that Russell released could not be put back in the bottle, and it would only get worse. In 1931, the Austrian mathematician Kurt Gödel (1906–1978) took another step. His brilliant contribution was to find a way to translate every mathematical statement about a number, including the proofs of such statements, into numbers. The procedure is strange but made something really amazing possible. It gives the numbers a voice and lets them start talking about themselves.

The consequences are far-reaching. To understand to what extent, we must use an even more powerful way of arguing, one that the mathematician Alan Turing (1912–1954) discovered a few years after Gödel. Turing was the one who more or less invented the computer, and the way he understood what Gödel had achieved was one of the first applications of his new idea. The key was to think in terms of computer programs. Turing imagined programs designed to calculate specific numbers, digit by digit. For example, there is, of course, a program that can calculate the decimals of $1/7 = 0.142857142857\ldots$, as far as desired. (I have just such a program on the computer I'm sitting at right now.) This particular sequence of numbers repeats itself and is not very exciting. There are also programs that spit out the decimals of $\pi = 3.1415926\ldots$, which is a much more interesting series of numbers without obvious structure. (Of course, I can also do this on my computer.) Number after number can be associated with appropriate computer programs that perform the calculations required to obtain them. Does this apply to all numbers? Let us see.

Turing imagined writing down a list of all kinds of computer programs and the numbers they produce. Strangely enough, he was then able to show that there must always be numbers that are not on the list. There must therefore be noncomputable numbers that no computer program can generate.

It is, of course, ironic that it was precisely Turing's computer that would lead to the discovery of numbers that cannot be calculated. It is also the case that such numbers are very common. In fact, the vast majority of numbers are noncomputable, while numbers such as 1/7 or π constitute the rare exceptions. This may come as a bit of a surprise, but the mathematical world is full of meaningless numbers that cannot be the outcome of any computer program and completely lack meaningful structure. What a waste! One can similarly complain, to no avail, about how our universe consists mostly of emptiness. Our planet hovering in the abyss of the universe can be compared to the number π surrounded by an infinity of noncomputable numbers.

How does this relate to Gödel? There is a threatening but interesting contradiction in the reasoning. Turing's argument was to construct a new noncomputable number with the help of his imaginary list of computer programs and computable numbers. But is not that the same as finding a method to actually calculate the noncomputable number? There must be a catch.

The bottom line is that there is no way to know if the programs on the list will really spit out any numbers at all. The calculations may just go on and on, without your ever knowing if they will stop or not. The list of programs and numbers will thus never be completed. Turing's crucial insight was that in general one can never know, nor can one ever prove, that a particular computer program will complete its computation. This is the "halting problem" that Turing proved to be unsolvable and that therefore also went against Hilbert's dream.

Back to Gödel. His next step was at least as strange as the first. With a formalism that could talk about itself, he could formulate a theorem that said that a certain theorem could not be proven. He could also make sure that the theorem that could not be proven was the theorem itself! So, what the theorem says is this: I cannot be proven! In other words, formalism acknowledges that it cannot meet Hilbert's demands. We who work outside of formalism realize that the statement must be true so that there can be no contradictions. If it were false, it follows that it must be provable. Of course, this cannot be true; it must not be possible to prove that a false statement is true. The only way out is that there must exist true claims that cannot be proven. Thus, no mathematical formalism can be complete.

An incomplete formalism can admittedly be supplemented, but it shows that there are choices to be made in mathematics and that what is intuitively true is about taste. As a consequence, mathematics cannot stand on its own two feet, independently of how we use it.

## Teaching Aliens Math

A common argument for the objective existence of mathematics is that it seems to be universal. In different cultures, mathematics itself has developed in different ways. The order in which different concepts were discovered (or invented) can vary. The areas of mathematics that have been in focus and considered important have shifted. Despite this, we are convinced that in the long run there can be no disagreement about what is true or false. It's just about formulating the ideas in a language that everyone can understand.

But if we were to encounter an extraterrestrial civilization? Could we be sure that their mathematics would be the same as ours? The problem would be about communication. Everything related to a common physical world would be possible for us to agree on, but what about the concepts we use in our thoughts? There may be choices of alternative ways of thinking that are difficult or impossible to communicate and explain. Despite this, many would probably argue that mathematics must be the same.

It is true that $1 + 1 = 2$ no matter who you are, and *pi* is $\pi$ even if you use a different symbol.

The fundamental problem is that we are limited to what we can accomplish by manipulating our common material world. We can use stones that we place in different patterns, rhythmic sounds, or flashing lights to try to impress with our mathematical ability. Whatever we do, it's about manipulating limited physical objects. It does not matter if it takes place on a computer screen. When we discover an underlying pattern, and succeed in setting up our own rules capable of reproducing what the other is doing, we can be endowed with the feeling of some kind of mutual understanding.

This is not necessarily easy. Think about how difficult it often is to make yourself understood in front of other people. As a university teacher, I often feel frustrated when I fail to convey an important idea in mathematics or physics to my students and discover that I am lacking the words. The best I can do is look prayerful and hope that they will understand in the end. To me, it's obvious, so why is it not to you? I am convinced that if I just manage to formulate it correctly, everyone will understand. Of course, the reverse is also true when I fail to understand students' explanations of what they do not understand. Why would it be easier with aliens—whether we are teachers or students?

I once tried to explain negative numbers to my then six-year-old son. Unable to directly convey my thoughts, I tried to find analogies based on shared experience. Money is an abstract concept that even young children can develop an understanding of. My son got a few dollars every week that he saved to buy a toy from time to time. I started with an example: "If a toy you want costs fifteen dollars, and you have only ten dollars, then five dollars is missing." He understood the point: He would not be able to buy it. "But if you borrow the five dollars that is missing, you will be able to buy it." Now he looked happy; there was hope! "But then you will owe me five dollars, which is another way of saying that you have minus five dollars." I took a break to let it sink in. "Ten minus fifteen equals minus five." Did he understand that? Perhaps.

What should we do if we completely lack a common background and experience with others? When we observe what the aliens are doing, we try to create models in our minds to understand what the aliens are trying to communicate. It does not differ much from what we do when we engage in physics in general. We study a physical phenomenon and try to construct a good model. The only difference is that in the case of the aliens, we suspect that there is another model, in the aliens' consciousness, expressed in terms of the mathematical concepts that the aliens use. Is it the same model we use? This is impossible for us to know. For those who have a religious view of the world, and also believe that the world is basically mathematical, the attempt to find out how the aliens think can be likened to revealing God's thoughts through studies of physics.

It's hard to even imagine how alien aliens could be. In my teens, I was a member of a local association for movie lovers. Every semester, some famous high-quality films were shown at the city library. The films were often difficult and not at all what was shown in the regular cinema. The adults in the club belonged to the intelligentsia in the city, and it felt special to be part of this group. One night, the film was Andrei Tarkovsky's *Solaris,* based on the novel by Stanislaw Lem. It is a science-fiction film set on a distant planet, where a strange sea plays an important role. Darkness fell over the theater and the film began. Suddenly, we were in an unknown world, where we hovered over a mysterious ocean. The image was enchanting and no sound could be heard. After a while, some people began to move around a little anxiously in their seats. Could there be something wrong with the sound? The film was restarted, now with music, but if you ask me, the experience was not at all as overwhelming.

*Solaris*'s thinking ocean shows how alien to us intelligence on another planet could be. What would we talk about in the event that we could communicate at all? Some people seem to think that music could be the key. As we contemplate a hopefully distant future when human civilization is gone, we can take comfort in the fact that the two *Voyager* probes on their journey out of the solar system carry with them all kinds of cultural treasures—including music by Johann Sebastian Bach. In the spring of 2020, *Voyager 1* had reached a distance of over twenty-one light-hours in the direction of the constellation of Ophiuchus. In a few tens of thousands of years, it will be among the stars several light-years away, and perhaps one day an alien civilization will find our priceless heritage. In any case, that is what we so desperately want to believe.

The taste of music varies over time and between cultures, although there are common themes. Perhaps music reflects the structure of our brains, and if other intelligent beings have something similar, it would be fundamentally different and impossible for us to understand and appreciate. It is likely that the beauty of Bach's music would go unnoticed and the aliens would have no understanding of the strange recordings at all. Even if they were able to build a record player properly, and perceive the sound, it would not matter. To understand music, one must have a human brain. When there are no human brains left in the universe, the first movement of the Brandenburg Concerto No. 2 will have finally disappeared from our universe, and the coded message on *Voyager*'s golden record will have become meaningless noise.

The point is that if human mathematics decisively depends on our biological nature and the construction of our brains and physical bodies, we are unable to understand how mathematics could be different.

The mathematics we use to model the world in the form of natural laws does not exist in the world itself. The laws of nature manifest themselves and are identical with physical patterns in our brains that reflect phenomena that we observe in the world around us. When the patterns are in tune with the world and we find consistency, we see the models as successful. Now do not get the impression that mathematics is a social construction. The mathematics we use is not arbitrary at all. But it lives neither in an extraterrestrial Platonic world of ideas nor in the external physical world independent of us. It exists purely physically in our biological brains and disappears with us and is in that sense a biological construction dependent on our biological nature. The laws of nature exist in our heads and are the tools we use to understand regularities in the surrounding world. What Max Tegmark and other hopeful Platonists do not want to realize is that all their dreams of a mathematical universe are formulated in a gray little lump of thinking brain matter. Mathematics exists only in the form of transient processes that help biological beings to better understand their enigmatic existence. The beautiful truths we find in mathematics, which make some people feel the presence of something almost supernatural, are only a consequence of our own limitations. With intellects significantly more powerful than ours, in brains with greater capacity, the deepest mathematical theorems would be seen as pure trivialities, comparable to counting $1 + 1 = 2$ on an abacus.

# 4

## There Is a Difference Between Model and Reality

*The world is large, very large. My head is small, quite small. It is impossible to put the world in my head. Nevertheless, we try to make some kind of representation in our body.*

—Jacques Dubochet,
Nobel laureate in chemistry,
speech at the Nobel Banquet, 2017.

At his Nobel lecture in 1988, the particle physicist Leon Lederman (1922–2018) told a story about an experimentalist and a theoretical physicist who are out on a mountain hike. They get lost and the theorist picks out a map from his backpack. He studies it carefully, looks up and points to a distant mountaintop, and exclaims triumphantly, "We are over there!"

The mistakes you make by confusing models with the world itself are usually a little more subtle. Nevertheless, it is often such mistakes that prevent progress in the understanding of new physics, and when they are corrected, the road toward great discoveries is cleared. The breakthroughs in physics often come when one questions the relationship between model and reality. One must identify the aspects of the models that do not correspond to anything real and trick us into believing what is really nothing more than fairy tales. Advances in understanding the world are not just about discovering new phenomena but also about driving away ghosts that were never real. Examples include Ptolemy's epicycles that Johannes Kepler killed, the phlogiston that Antoine Lavoisier (1743–1794) replaced with oxygen, and the ether that Albert Einstein cleaned up when formulating his theory of relativity. The absolute time that Newton introduced turned out to be a flawed mathematical construction with no equivalent in the real world. Through the theory of relativity, Einstein was able to unite space and time into a unit, space-time, where time was not absolute but was determined by the observer's motion.

Constructing models is not something that is unique to science. In everyday life, we constantly make use of different models. Some are subconscious and are more or less embedded in our brains and our nervous systems. When we walk, run, or even lift a glass of water to drink, inner representations of how the objects move are secretly manipulated, leading to results that are then translated into decisions about how to tense or relax muscles to achieve what we strive for. Playing soccer, or just throwing a stone at a specific target, requires the use of even more sophisticated models with ballistic trajectories.

It's not just people who build models. For all life, whether it is a butterfly trying to land on a flower or a root seeking nourishment in the soil, it is about creating models of the environment. It does not matter if what lies behind this is instinct or conscious decision. It's all about models anyway. Creating models that represent relevant phenomena in the surrounding world in order to be able to adapt and survive in this way is simply what living organisms are engaged in. Scientific model building goes a step further by allowing formal mathematical systems to reflect the natural, real world. Cause and effect are coded into mathematical relationships and are thus elevated to natural laws. Early observations of how celestial bodies such as the Sun, Moon, and planets moved, and the regularities discovered, suggested that mathematical model building could be a successful path.

In order to really be able to make predictions, it is required that we not only know the laws of nature but also something about what the system we want to describe looks like at a given moment—for example, right now. These initial conditions are translated into a mathematical language and used as input to the mathematical laws. A calculation transfers input to output and we interpret the result as a prediction of what will happen. Hopefully, the result agrees with what we can then measure.

In short, this is how we have looked at science since Newton's time. This is in some sense what science is.

## What Is Real?

I'm a realist. I sincerely believe that the world exists independently of my own existence and that there are truths about the world that I can try to reveal. Some of what I already think I know, I'm convinced is correct. Other things, and I'm not really sure what, will turn out to be wrong. This does not worry me so much. I'm a scientist. I'm used to being wrong, and I have to admit that most of the ideas I've had during my career—especially the most interesting ones—have proved to be useless. The search for the truth is full of traps.

There are many different kinds of realism. A particularly popular version among scientists claims that there is a world outside our consciousness that is completely independent of our ideas and preconceived notions. There is one and only one way in which the world really is. This is a belief that many scientists profess, and isn't this exactly what realism must mean? Can there be any alternatives? There actually are alternatives, and that is why this particular type of realism has been given its own name: *metaphysical realism*. But this is not a position to be taken lightly. It has far-reaching consequences that go against our intuition and how we look at and relate to the reality of everyday life.

If there is only one way in which the world truly is, it must be in the form of a collection of fundamental particles that physics claims build the world. Metaphysical realism allows no alternative ways of looking at reality. The all-encompassing description of fundamental physics is exclusively exclusive. All macroscopic objects that we see around us are revealed as nothing more than arbitrary constructions. This includes not only the chair where I am sitting right now and the chair where you are sitting (if you are sitting now), as well as the computer where I am writing *The World Itself* and the book you are holding in your hands, but also my body as well as yours. None of this exists according to metaphysical realism. The world we think we are in is an illusion that we ourselves have invented. It is only thanks to the very latest achievements of physics that we have finally managed to reveal the true nature of reality and identified the only thing we need to relate to from now on.

The philosopher Hilary Putnam considered such a view to be pure madness. After all, this is exactly what many who think they understand physics want to convince you of. Honestly, I have tried to tell the same story many times, but eventually I have changed my mind.

If one does not want to confess to metaphysical realism,

one can fall for the temptation to simply deny realism and claim that there is no objective reality out there at all. It's all about arbitrary constructions that are influenced by the culture in which you find yourself and how you grew up. We call this *attitude relativism*. For the relativist, fundamental physics also plays by the same rules and therefore does not hold any privileged position. Of course, relativism has its own dangers. While the metaphysical realist may become depressed at the thought of being just a heap of particles, a pure-blooded relativist (if there are any) risks more immediate dangers, such as falling off cliff edges or being hit by cars claimed to be nothing more than social structures.

Neither metaphysical realism nor social constructivism seems to be a very good way to understand the world. Is there any philosophical model that recognizes both an objective world and the possibility of making subjective images of it? There is, and Putnam gave it a name: *internal realism*. The point of internal realism is that it accepts an objective world out there, but the way you can make it comprehensible is not unique. Let us start from, and embroider on, Putnam's own example of a world made up of only three kinds of objects. For simplicity, we'll call them apple, orange, and banana. These are the only building blocks we have at our disposal. It seems to be an indisputable fact that the world consists of only three types of objects. Right? Isn't that exactly how this little world is defined? Not necessarily. There is nothing to prevent us from picking the fruits together in pairs, or even triplets, and considering these as additional building blocks. If we reason in that way, we have as many as seven different objects to work with. In addition to apple, orange, and banana, we have apple – orange, apple – banana, orange – banana, and apple – orange – banana. Everything that happens in the world can be expressed and made comprehensible with this scheme. Are there three or seven different building blocks? It is up to you. But if you choose a scheme, and stick to it, there are certain statements about the world that are objectively true.

It works in a similar way when we try to get a grip on the real world. The world itself can be made comprehensible by using the building blocks of fundamental physics such as atoms and voids, or everyday objects such as chairs, books, and people. It is in a fundamental sense the same world—we do not deny realism—but internally, in our consciousness, we can choose to think of the world in different ways. Surely it's elegant.

Is one way better than another? Sure, but it depends on who you are, where you are, and what purpose you have. There are definitely differences between how various people conceptualize the world. If you take a course in physics, you will be introduced to new and perhaps foreign ways of looking at the world. You realize that there are small and invisible particles that act in the innermost part of matter, and when you look up at the night sky, you see something more than just flashing dots of light. If you, like me at an early age, became obsessed with physics and science, you will be surprised that not all people see the world in this way—even if they are highly educated in other subjects. Despite these differences, all of us humans still have a lot in common. Many models of reality that we use are genetically inherited and deeply rooted in our brains and bodies, or constructed based on experiences from daily life, and in many ways the same, even if they are not identical. If this had not been the case, we would not have been able to communicate with one another and evolution would have been one great failure.

The advantage of internal realism is that it differentiates between the real, existing world and how it is described. The goal of science is to offer efficient and reliable models that can be used to make useful predictions about the world itself. Each scientific model is preliminary and there is always room for improvement—not only in terms of quantitative predictions but also in terms of conceptual substantiation. The step from Newtonian mechanics to the general theory of relativity is, as we have seen, an example. Newton realized how the fall of the apple and the Moon's orbit around the Earth were controlled by gravitational forces. Einstein turned everything upside down by denying that there were any gravitational forces at all and instead claiming that everything was a consequence of curved space-time. This kind of conceptual revolution can give the impression that science is making progress by tearing down everything it has built up, only to start all over. At least if you are a metaphysical realist: Back to square one after being wrong again. Over and over again. Newton turned out to be wrong and Einstein took his place. In other words, one can never trust science. An internal realist sees it differently and notes that science works exactly as it should.

At this stage, it may be practical to introduce the terms *ontology* and *epistemology*. Roughly speaking, ontology is about what really exists. For real . . . *das Ding an sich,* if we cite Kant. Epistemology, on the other hand, is about what we can really know and is a much more practically oriented field, closer to what science is really about. You can test claims about what you measure, but you will never really know what is going on behind the scenes.

From a scientific point of view, one may be tempted to claim that ontology is meaningless and that you only need to care about what can be measured. If only it were that simple. When we think of the world, when we use our experience and creativity to develop our understanding, it is impossible not to have, somewhere in the back of our minds, a concept of an independent existing reality that we are trying to get to know. It would be unsatisfactory not to have it—if one does not now want to fall back into full-scale relativism. The point of internal realism is that it accepts the existence of a real world out there, the world itself, where everything is about ontology. What we do as scientists, humans, or living beings in general is to struggle with what we have direct access to—that is, models of the world. We cannot place the world inside our little heads, but we can try to represent it as best we can.

## Models for Real

The difference between a model and the world itself is profound but often difficult to maintain in practice. Physicists often discuss their models of reality in a way that can be confusing and misleading. Mathematical constructions that are no more than intermediary tools, without experimental significance, are mistaken for being real due to their formal role in a successful theory. Many physicists in this sense have a rather naïve view of the world.

Sometimes the models are so well supported by experiments and observations that it is justified to take the leap fully and claim that the mathematical objects can be identified with something that actually exists. An illustrative example comes from particle physics. The concept of quarks was introduced in the 1960s by the American physicist and Nobel laureate Murray Gell-Mann (1929–2019), and independently by the Russian-American physicist George Zweig, in order to conveniently describe the patterns that could be discerned in the confusing number of new particles found in addition to familiar particles such as protons, neutrons, and electrons. It worked in a wonderfully beautiful way—no one could deny the power of the mathematics used. Despite this, it was not very modern at the time to imagine that there were some building blocks that were more fundamental than others. It felt old-fashioned and was a little too reminiscent of ancient Greece. Had we really not reached a greater maturity? Many physicists expected something much more sophisticated and democratic, where all the particles that the experiments revealed in some way consisted of one another. They had nurtured a beautiful thought, which turned out to be wrong. And it would be a while before the quarks were fully accepted. Sure, you could pretend that they existed when you performed your calculations, but in reality they did not exist. The critics finally gave up and agreed that what smells like quarks, tastes like quarks, looks like quarks, probably also are quarks. Further experiments made it impossible not to accept them as objective constituents of matter. Nowadays, all physicists embrace quarks and all other particles in the standard model of particle physics. There are no major conceptual problems to worry about, and at the time when I write this, the model and reality seem to agree within the allowed errors.

Another much older example is the controversy between the Catholic Church and astronomers over the choice between heliocentric and geocentric systems. The ideas had been circulating for thousands of years, but it was the Polish astronomer Nicolaus Copernicus (1473–1543) who managed to deliver the news that it was best to view the Sun as the center of the solar system. The Church was not overjoyed but saw the point of the simplifications that followed. A lot of calculations actually became much easier to perform if the Sun was placed in the middle. It was free for anyone to follow Copernicus in calculating the motions of the planets—if it became easier that way—but in reality, it was of course the Sun, not the Earth, that moved. Just as hesitant physicists eventually had to accept the existence of quarks, the Church eventually had to acknowledge its mistake concerning the motion of the Earth. Formally, this happened in 1992, when Pope John Paul II rectified the Church's previous position.

The physicists of the 1960s, as well as the priests of the sixteenth century, definitely cared about how the world was set up. Not just about how to describe it. The underlying ontology is a source of inspiration and makes a scientific endeavor something more than a formal game. For the individual researcher, it is important that what one believes in is also true, even if the science itself at a given stage can be agnostic and experiment with competing ontologies that give the same predictions.

How we choose ontology can also be decisive for how science progresses. In a given situation, it may not matter so much for the predictions we make exactly how we look at specific components in our models. Do they respond to something real or not? The calculations give the same results and the predictions we can test are identical. But sometimes, and this is illustrated by the examples I have given, it turns out that the ontologies in question, on closer inspection, differ in their predictions. One way of looking at reality is a dead end, while another leads to new discoveries.

## The Theorem of Löwenheim and Skolem

Neither the model of the world that fits in a butterfly's tiny brain nor our scientific theories, no matter how clever we think they are, can be identified with the world itself. There is a difference between model and reality that can never be ignored, but as long as science is about an external world far from the human observer, one can afford to be a little careless with this fact. As a physicist, you can stick to a rather naïve picture of what physics is really about. It is a common misconception that theoretical physicists with expertise in areas such as string theory and cosmology are the ones best placed to lay out the foundation of existence. Not even if they get help from the more experienced in the field, the analytical philosophers, do they always end up in the right. As we shall see, the analytical philosophers are also at risk of the same kind of naïveté.

Let us see how attempts have been made in physics to clarify the important distinction between reality and model. On the one hand, it is believed that there is an objective world out there, independent of us, which is best described by various "set theoretical constructions." This is just a fancy way of saying that the world is made up of a lot of things. For example, apples, oranges, bananas, and maybe fruit salad. On the other hand, there are scientific theories in the form of mathematical and logical axioms made up of sequences of mathematical symbols. Meaning is considered to arise when the symbols are connected to the real world and thus are interpreted in a certain way. The claim is that this is all you need to keep the world apart from your image of it.

Unfortunately, it is not that simple. Throughout the history of physics, there are several examples of what we actually have to deal with. The French physicist and philosopher of science Pierre Duhem (1861–1916) argued that we can never test a scientific hypothesis in isolation, but only together with other hypotheses. If the prediction of a certain hypothesis turns out to be wrong, we do not know if the hypothesis itself is wrong or if the error lies in some other assumptions we have made. An excellent example is the discovery of new planets in the solar system. In 1783, the British astronomer William Herschel (1738–1822) discovered by chance Uranus, the first planet to be discovered since prehistoric times. Astronomers carefully studied its motion, and Newton's mechanics were used to characterize the orbit. Surprisingly, something was not quite right. The planet seemed to violate the law of gravity, and its measured position insisted on deviating from the predictions. What was the reason? Assuming that Newton's laws actually applied, the French astronomer Urbain Le Verrier (1811–1877) and the British astronomer John Couch Adams (1819–1892) predicted the existence of a new planet. The German astronomer Johann Gottfried Galle (1812–1910) discovered Neptune in 1846 in just the right place to explain the deviations in Uranus's orbit. In a similar way, Mercury exhibited strange behavior, which inspired Le Verrier to proclaim the existence of another new planet, Vulcan, near the Sun. This time, the outcome was different. No planet was found and the fault turned out to be Newton's. In 1915, when Einstein put the finishing touches on his new theory of general relativity, he was able to show that Mercury moved exactly as it should and there was no need for a planet Vulcan.

Strangely enough, Neptune did not seem to move in the right way, either, which led Percival Lowell (1855–1916), among others, to predict the existence of a large planet farther out from the Sun. Incredibly, history seemed to repeat itself when Clyde Tombaugh (1906–1997) in 1930 discovered an object in just the right place in the sky. The new planet, Pluto, turned out to be a bit of a disappointment. It was far too small—we now call it a "dwarf planet"—and could have no measurable effect on Neptune's orbit. It just happened to be in the right place by chance. After careful measurements, it also turned out that there were no unexplained oddities in Neptune's orbit.

We can thus state how three different planets seemed to move in unexpected ways with three completely different explanations. Science is full of moving parts in a network of interconnected hypotheses and ideas. Everything must be tested at the same time, which is a strength as well as a weakness. The conclusion is natural in science, but the analytical philosopher Willard Quine realized that it also applies to language in general.

A key to staying on the right side of the relationship between model and reality is what is usually called the *Löwenheim-Skolem theorem*. Leopold Löwenheim (1878–1957) was a German mathematician who gave the first proof of the theorem in 1915, while the Norwegian mathematician Thoralf Skolem (1887–1963) implemented a significant simplification five years later. The theorem is certainly quite technical, but it has far-reaching consequences for our reasoning. Basically, the theorem claims that there is no easy way to relate reality and model. Whatever you try to describe can always be identified with the integers, regardless of your intentions. In other words, *everything you can talk about can also be counted.* Whether you are trying to describe what you know about cosmology, the view from your window, your political views, or the love of your children, it can always be misunderstood by those who do not speak your language as a statement about integers. Pretty disappointing. A language with its rules, grammar, and symbols is in itself completely meaningless. If no one tells you how language connects to an external world, the only thing you can do is map the logical structure of language. Everything degenerates into pure and rather uninteresting mathematics that does not say anything about what the intended message really is.

A couple of years after Skolem finished his proof, he noted that it also followed *that what you cannot count, you cannot talk about, either.* That statement goes by the name *Skolem's paradox.* The reason why this seems paradoxical is that there is a lot you cannot count that you still can talk a lot about. In mathematics, it is popular to talk about so-called *nonenumerable infinities.* The German mathematician Georg Cantor (1845–1918) constructed and classified just such infinities. Hence, there are infinities that are greater than other infinities. The number of integers is infinite, but you can count them: 1, 2, 3, etc. You can go on for as long as you like, but you will not miss any of them. When it comes to real numbers, it's a different matter. They are not enumerable. You can try to create an infinitely long list of all real numbers between 0 and 1, where you number them 1, 2, 3, all the way to infinity, but no matter how you do it, you will always miss some of the real numbers. There are always new numbers you'll need to add. How could Cantor—and we right now—talk about such numbers, and explain their characteristics, if Löwenheim and Skolem claimed it was impossible? Cantor actually lost his mind, and mathematicians generally do not seem to care too much about the problem and happily continue to talk about different kinds of infinities. Like the rest of us, mathematicians are not only interested in syntax and symbols but load language with semantics and meaning. Mathematicians are not at all as pure-minded as one might think.

For linguistics, this was a severe blow. Hilary Putnam went further and used the theorem to show that a language cannot determine its interpretation on its own. This applies not only to mathematics but to language and thought in general. A language isolated in itself, regardless of the number of words and the grammatical structure, is completely meaningless. What I now sit and write on my computer, or the words I shout to the family, make no sense in isolation. Everything could just as easily be uninteresting statements about integers. According to Putnam, there is no way to capture the world in a meaningful way just by talking. Our universe contains so much more than can be captured by integers. Not even if our universe corresponded to mathematics in the form of no more than real numbers would it be possible to capture it with the help of axioms. Whatever we formulate in words could be realized by integers.

Putnam argued that there is no unique way to connect words and statements to the condition of the world. It follows that there are no analytical claims, either. Nothing in itself is obvious. Not even the classic example "all bachelors are unmarried" can be considered without reference to something else. Löwenheim-Skolem draws our attention to the fact that if you want to figure out what "bachelor" and "unmarried" really mean, you must also know a lot about the language and how it relates to the world. How else can we be sure that bachelors really have to be unmarried? It follows that there are no isolated truths that we can completely rely on. Even more worrying is that science itself, through a similar reasoning, becomes impossible. All the theories we put forward, no matter how well they seem to reproduce reality, are only preliminary constructions that can at any time fail and require replacement by something else. The paradigm shifts in science can be likened to suddenly discovering a bachelor who is married. The conclusion is devastating and depressing. There is no secure knowledge and everything is relative. What a shame.

Is there a way out? Certainly. But first we will see how

Putnam-Löwenheim-Skolem go wild with some other popular ideas about human language.

## The Embodied Language

It is strange how the notion that everything is mathematics has remained so popular for millennia, in its various disguises. It is a completely unreasonable idea that in practice has a religious basis. And when we have now seen how far from obvious the relationship between language and reality is, and how the same difficulties also affect scientific model building with the help of mathematics, it becomes absurd to see mathematics as independent of us. It is, of course, a bit ironic that it is with precisely a mathematical reasoning that one can show that the world cannot be mathematics. An even more direct consequence, to which I shall return later, is that a computer, which is nothing more than an implementation of pure formalism, in itself cannot carry meaning and reasonably cannot be conscious, either. The crucial problem is the common notion of free-floating concepts without physical anchoring. The disturbing relativism that follows from Löwenheim-Skolem can only be tamed if we define all concepts based on our bodies and brains. There are no abstract and formal symbols in the brain tissue, or anywhere else, for that matter, with a mysterious relation to a higher and independent world of mathematical concepts. Instead, there is a match between purely physical phenomena inside and outside our bodies.

Funnily enough, there is a parallel history in linguistics. But instead of trying to understand the origin of the universe or life, linguistics tries to figure out what defines language, how we acquire it, and how we can trace its origin. One of the greatest linguists of our time is Noam Chomsky. He stands firmly with his feet in the formalist tradition and sees the relationship between language and reality precisely in the way I have criticized. According to Chomsky, a language is also absolutely necessary for human thinking.

Chomsky encounters far greater difficulties in the origin of the human language than the physicist does in the role of mathematics. The physicist is content with the fact that mathematics works and delivers at most rather sweeping statements about how surprisingly efficient it is. If the physicist is pressured, especially a theoretical physicist, there is a risk that he or she will take a position similar to Max Tegmark's, and identify mathematics with the world itself. Fortunately, there is no reason to take such claims seriously. The widespread speculation is difficult to test and can be viewed as pure entertainment. Chomsky has greater demands on him. The role and origin of language is a respected scientific field and one has reason to expect clear and unambiguous answers. He cannot hide in the same way as someone who engages in speculative physics.

To defend his position, Chomsky is forced to draw a number of conclusions that must be seen for what they are: falsifiable predictions. He claims that language is unique to humans and that no other animal has anything like it. It is not about degrees but about an absolute and qualitative difference between the abilities people possess and those that chimpanzees or dolphins master. According to Chomsky, human language has not evolved gradually from anything that existed before man. But where does it come from? Chomsky does not believe in any intervention by God or aliens, but concludes that the explanation is a random genetic mutation. All people have a universal ability to understand and use language that enables us to connect with the world of ideas thanks to this historical fluke.

It is an extraordinary statement of metaphysical propor-
tions, and yet a popular idea for a simple reason. It helps us
draw a sharp line between us as the crown of creation and
all the other, lower beings. It is an attempt to maintain a
dualistic view of the world that protects us from becoming
just another kind of animal. If God is not responsible for the
miracle, it may as well be chance. All this is compatible with
the notion of a Platonic world of thought that contains not
only mathematics but also languages with which we have
succeeded to establish contact.

But research on animal behavior points in a completely
different direction. Biologists have discovered that chimpanzees have sophisticated sign language, while sperm
whales sing complex and culturally transmitted songs.
Israeli biologist Yossi Yovel has developed a computer program that can decipher what Egyptian fruit bats say to one
another. It's mostly about fussing, in the manner of "Stay
away" and "Do not wake me." A strange empirical law,
Zipf's law, named after the American linguist George Zipf
(1902–1950), seems to work just as well for the whistling of
dolphins as for all human languages. According to Zipf's
law, the relative number of times a certain word is used is
$1/n$, where $n = 1$ corresponds to the most common word.
According to Zipf's law, the second most common word is
used half as many times as the most common, and so on.
No one knows why this is so, but apparently it applies just as
well to English as to the sounds dolphins make.

There is also genetic evidence that we have a lot in common with other animals when it comes to language. In the
1990s, researchers discovered that defects in a gene called
FOXP2 could be linked to seven different speech defects.
The gene is found in many different vertebrates, and allows

birds to sing. It is likely that there are many genes that affect our ability to speak and that also have many other functions.

Chomsky argues that human language differs from the language of other animals by being *recursive*. This means that it can be "nested" as far as you like—even if it eventually becomes difficult for us to keep up. A nested sentence contains a clause that is embedded into another clause. If I write "I wrote this sentence," in response you can reply, "I read the sentence that you claim you wrote." And so on. According to Chomsky, animals are incapable of using language in this way. This is, however, far from certain and has also been questioned. In 2011, researchers at Kyoto University were able to show that Bengal finches have a recursive ability. The birds sing varied songs that are built from a fixed number of syllables, and they are sensitive to whether what they hear is grammatically correct. By exposing the finches to different sequences of syllables and then studying their reactions, it turned out that they could handle nested sentences. It was also possible to locate this ability to a specific place in their brains. Despite this, it is not clear if this is just syntactic showing off or if the songs carry a semantic meaning.

We have already seen the problems that arise when we try to match the world against abstract symbols and from this obtain meaning. It simply does not work. This is also the reason why computers, at least of the kind we can now imagine, cannot think. They use pure syntax and the only meaning that is there is the one we ourselves project. The alternative is about an embodiment where thoughts are more than just a mechanical manipulation of meaningless symbols. The American philosophers Mark Johnson and George Lakoff argue in their *Philosophy in the Flesh* that language and mathematics are based on our physical bodies and thus generate meaning. Here we can see a way out

of the difficult dilemma that Löwenheim and Skolem discovered: Information for itself, without a material manifestation, is nothing. Information without matter simply
does not exist. It is matter that matters.

Let me give an example: a road sign with the number 70.
The sign informs us that here you must not drive faster than
70 km/h (at least if the sign is in Sweden). This information
is not at all a property of the sign itself. There is not even
a "70" in the matter that makes up the sign. If a symbolic
reform was carried out where the numbers 3 and 7 changed
meaning, this would in no way change the character of the
matter of which the sign consists. In isolation, it does not
contain any information about any speed limit. In order for
what the sign is intended to convey to have any meaning,
another material system is required, usually the brain of a
car driver, which makes an interpretation of the sign and
hopefully adjusts its behavior. The relevant information
does not exist independently of this interaction, but only in
relation to an interpreter.

The point is that science, when seen only as a system based on mathematical logic, has no meaning. What
researchers like myself do in our theories is to manipulate
symbols according to formal rules. It is only when these
symbols are connected to the real world, or, more precisely,
the aspects that we select and abstract, that meaning is gen-
erated. The problem is that there are crucial steps, which
are mistakenly considered trivial and deliberately ignored.
Between the high-flying ideas and the messy natural world,
which is what science is all about, lies the embodied consciousness of the researcher himself. There is no objective,
external, and independent connection between the abstract
world of mathematics and logic and the universe. The connection is always made in a brain of flesh and blood.

# 5

## Computers Are Not Conscious

*A pair of wings, a different mode of breathing,*
*which would enable us to traverse infinite space,*
*would in no way help us, for, if we visited Mars or*
*Venus keeping the same senses, they would clothe in*
*the same aspect as the things of the earth everything*
*that we should be capable of seeing. The only true*
*voyage of discovery, the only fountain of Eternal*
*Youth, would be not to visit strange lands but to*
*possess other eyes, to behold the universe through*
*the eyes of another, of a hundred others, to behold*
*the hundred universes that each of them beholds,*
*that each of them is . . .*
     —Remembrance of Things Past: The Captive,
     MARCEL PROUST (TR. C. K. SCOTT MONCRIEFF)


SWEDEN'S MOST IMPORTANT contribution to the early
history of philosophy was to kill Descartes—the great
philosopher who looked inward and became convinced that
he existed because he thought. He not only stated that he
was a thinking thing but that this was all that mattered. He
did not deny that he had a body but said that "the mind is
really distinct from the body and can exist without it."
     Descartes, the acclaimed genius who was about to lay

the foundations for a new era of rational thinking in human history, made a fatal mistake. Several years after he made the above declaration, he was tricked into going to Sweden by the young Queen Kristina, daughter of King Gustav II Adolf, who through wars had made Sweden a superpower at that time. Sweden had a great need to improve its reputation by acquiring some culture. Kristina herself was genuinely interested in science and philosophy and had aroused the interest of the great philosopher through her letters. The poor Frenchman arrived in Stockholm, on the edge of the civilized world, in the fall of 1649 to tutor the queen in philosophy. He was accustomed to working late and sleeping long into the morning, but he was forced to get up early to give lessons to the queen in the cold and damp castle. His argument that the soul can exist independently of the body must have appealed to the deeply believing Kristina, but his mechanistic philosophy in general was probably less popular. He soon contracted a cold, which developed into pneumonia, and died in miserable condition in February 1650, just a few months after his arrival.

He was buried on the outskirts of Stockholm, where he remained for sixteen years, until France decided it was time to bring home the remains of its national hero. He was buried a second time in the church of Sainte-Geneviève in Paris, where he would rest for more than a hundred years while the church slowly fell into ruin. During the French Revolution, his remains were rescued from the dilapidated church and temporarily stored in the Musée des monuments français. There he would have to wait until 1819 before receiving his final resting place in the Abbey of Saint-Germain-des-Prés—without his skull. As early as 1666, before Descartes's body had even left Sweden, the skull had been stolen by the captain in charge of guarding his coffin. The skull was then housed by a long line of scholars before eventually ending up with Linnaeus's disciple Anders Sparrman, who had accompanied James Cook on his second voyage to explore the southern oceans. The French Academy of Sciences followed the trail and found that the skull was genuine. What about the rest of the body that was allegedly saved from the ruins of the church of Sainte-Geneviève? The bones had been found in a wooden coffin, even though they should have been in a copper one. It would seem that Descartes's body had been neglected, while his skull had been rescued by a thief. And here it is, covered with notes made by its owners over the years.

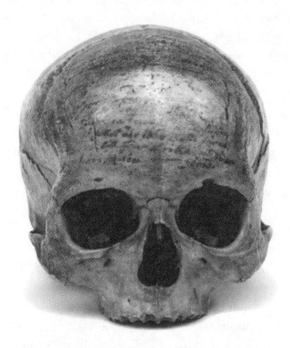

*Photo of Descartes's skull by J. C. Domenech,*
*courtesy of the Muséum national d'histoire naturelle*

It is in this hollow piece of bone that Descartes's consciousness observed itself and concluded that it existed and was thus independent of the body. The poor body lost, the only thing left is the empty container that once contained his brain. But where has his consciousness gone?

## Body and Soul

The struggle to sort out the relationship between consciousness and body has continued for millennia. Different cultures have drawn various conclusions and come up with one imaginative proposal after another, not only about what happens after death but also about where consciousness is actually located. A remarkable example is that of the ancient Egyptians, who were convinced that the physical body played an important role even after death. They located consciousness not in the brain, which was considered a rather useless filling of the skull, but in the heart.

Nowadays, we are so convinced that we think with our brains that we tend to forget that we have physical bodies. We may have our doubts about what this "I" or "myself" really is, but we are convinced that it is inside the skull. The Egyptians believed the human soul consisted of two key components, Ka and Ba. Ka was a kind of spiritual equivalent to the physical body that left the body when it died and hung around by the grave. Ba was more like a personal soul that after death could travel freely in this world and others. To make this possible, the physical body had to be preserved so that it could offer the soul a safe haven to return to at night. If the body fell apart, Ba and Ka got lost, leading to an undesirable end. With a bit of goodwill, one can interpret this as a notion not unlike that of the devoted physicalist who believes that consciousness cannot exist independently of the physical body. These issues that the ancient Egyptians wrestled with are still highly topical today.

The dualism that Descartes advocated was a decent reflection of what many religious people still professed and was thus not that original. But it gave a philosophical perspective on the issue and made the conclusion seemingly inevitable, regardless of belief. Nowadays, this separation between body and soul is not very popular among the scientifically educated. At least that's what they say. Many would argue that consciousness is rooted in matter and cannot exist independently of it. A free-floating self without anchoring in a physical organ like the brain cannot exist. Despite this, the old Cartesian dualism retains its allure and manages to infiltrate the thinking of modern computer science. The way in which we distinguish between hardware and software is strikingly similar to how we used to look at body and soul. In this way, computer technology has developed into something not only for those who are interested in facilitating heavy calculations but also for those who ponder the question of Descartes's "I" in a new guise.

## How to Play Chess

When I was in high school in the early 1980s, I wrote a simple computer program in the programming language BASIC that played chess. My strategy was simple. I defined a table of numbers that represented the positions of all the pieces on the board and listed all the rules for how the pieces could be moved around. The program included a routine that tested all possibilities a few moves forward. It was difficult to decide what was a good move, but I chose the simplest option. I simply let the program associate a numeric value to each piece, such as three points for a bishop and five points for a rook. The program then chose the move that—given that the opponent chose his or her best move according to the same principle—led to the highest score. The principle is used by many of the most advanced chess programs and is called *minimax*.

I remember the long printouts that I carefully scanned for the slightest mistake in my attempts to improve the primitive algorithm. The program I invented, and which eventually worked, was not a particularly skilled player, but it could at least find an available checkmate in one or even two moves. It was good enough to beat a beginner and at least annoy someone with a little more experience.

It never occurred to me that my program, which fitted on a few pages of paper, or the computer that ran it, would be conscious or intelligent in any sense. I fully understood how the program worked and realized that it only consisted of a collection of rigid and deficient rules that, if executed in the right order, calculated what the next move would be. It was a purely mechanical process that in principle could have been implemented with the help of a machine consisting of gears and rods. But it was fun that it worked. As a university student, I translated the program into a slightly more modern language, improved it a little, and then abandoned the entire project.

Nowadays, you can download chess software on your phone that beats any human player. The first time a reigning chess champion was defeated by a computer in competitionlike conditions was on February 10, 1996, when Deep Blue, playing as white, forced Gary Kasparov to give up after thirty-seven moves. It was a revolutionary moment, not just for Kasparov, as he describes in his book. Although he won the overall match, which consisted of a total of six games, the following year an updated Deep Blue took revenge and it was clear who—or rather, what—was the champion. Finally, a human had been outwitted by a machine in a game that is considered to be an art form.

When I try to play against a computer that defies all my plans and delivers one destructive move after another to finally humiliate me, I am emotionally torn. Could it be that I am in the presence of superhuman intelligence? Chess is not the only game where a computer has triumphed over man. The human ego was defeated once again when computers learned to master the ancient Chinese game go. The rules for go are much simpler than those for chess. You play it on a board of 19 x 19 lines (there are also other versions). One of the players has white stones, the other black, and each takes turns laying stones where the lines intersect. The goal is to surround the other player's stones, which are then removed, and finally conquer the entire board. It may seem simple but go beats chess in complexity: The number of possible moves has been estimated at more than $10^{170}$.

The mathematician and grandmaster of chess Emanuel Lasker (1868–1941) once exclaimed, "Chess is a game that is limited to this world, Go has something extraterrestrial about it. If we ever find an extraterrestrial civilization that plays a game that we also play, it will be Go, without a doubt." There is something to this. The rules of chess are arbitrary; go is a completely different thing. In addition to the size of the board, which does not matter much, the rules almost write themselves.

It took a little longer for computers to learn to master go, but it was clear that it would happen sooner or later. The AlphaGo program, developed by Google, was the first program to beat a high-ranking player when it defeated Korean Lee Sedol in March 2016. Like all chess programs until then, it was preprogrammed with tricks and strategies used by skilled players, along with large libraries of sequences of possible opening moves. In this way, the program relied on what human players came up with in combination with a superior ability to count that made it possible to see many moves ahead.

At the end of 2017, something even more remarkable happened. A new version of the program, AlphaGo Zero, was developed that used a completely different kind of strategy. It had not been provided with any prior knowledge, or prejudices, about how to play good go—apart from the basic rules. Instead, it developed its own strategies by playing against itself over and over again. After only three days of training, and close to five million games, it was stronger than any human player and it soon thereafter became better than any other program.

Before we get into what this really means, let's go back to chess. An upgrade of the program, AlphaZero, could not only train to become a master of go but perform the same feat in chess. After just four hours of training, all on its own, it had learned enough to beat all the best chess programs. In December 2017, it played one hundred games against the reigning champion program Stockfish. It won twenty-eight times, got a draw seventy-two times, and did not lose a single time. During its training, AlphaZero had discovered all the chess openings that humans have designed and, in a sense, verified what it took a century, and several generations of human players, to discover. As it tried different possibilities, it started to play some a little less often than others and give the impression that it thought them less successful than some human players thought.

In this way, chess gained a new meaning independently of the people who until very recently were masters of the field. Far from being arbitrary and possibly artistic choices, up to the players' tastes, the chess openings took on the character of mathematical structures. The conclusions about how to best play chess that AlphaZero has come to are thus to be equated with objective mathematical results. This would suggest that some of the openings that AlphaZero has found incompetent should be provided with warning signs in chess manuals.

In addition to a strange love for the queen's pawn, AlphaZero is laid-back when it comes to sacrificing pieces. Unlike my old program, which only valued a position based on the value of the pieces, AlphaZero is only focused on winning. Long-term benefits in the positions are what counts and a piece that is trapped does no good. All this became apparent in AlphaZero's games against Stockfish. Experts who analyzed the game were amazed.

Human players, as well as chess programs, whether they are champions or primitive programs like the one I wrote in my youth, like to take pieces and hate to lose them. The strategy generally works well and is easy to implement. Games between beginners often boil down to a fight, especially for the queens. If you lose the queen, the game is generally lost. Experienced players learn to read the benefits of how the pieces are positioned and can dare to sacrifice a pawn or two. AlphaZero lacks human prejudice and simply searches for a way to win, adopting strategies that people find inconceivable.

Interestingly, AlphaZero is not very fast when it comes to processing the number of positions. It can only handle around eighty thousand positions per second, compared to the staggering seventy million of Stockfish. You could say that it is a thousand times slower than Stockfish and plays more like a human—but a very strange human.

Will all chess masters in the future focus on the queen's pawn? One possibility is, of course, that the human brain does not have the ability to implement the advanced strategies that AlphaZero discovered. It may be that they are too difficult for us to master. I have to admit that I find all this a little scary. Are we in the presence of a superhuman intellect with which man cannot compete? AlphaZero uses neural networks. These are algorithms that mimic how the human brain works. Neurons send signals to other neurons depending on which signals they, in turn, register. But what determines whether they should send a signal, and how strongly they should signal, is not fixed, but changes over time. This is exactly what learning is all about. The networks are universal and one can easily imagine how the abilities could be applied to other problems and not just innocent games. Could they be applied to science? Absolutely. This is already under way in areas such as protein folding and astronomy.

My own work as a theoretical physicist sometimes feels a bit like playing chess. I am always looking for patterns and opportunities to follow where they lead. Sometimes success comes—checkmate!—but it more often leads to humiliating defeat, where I am forced to admit that I was on the wrong track. A small typo, a move I did not anticipate, and everything falls apart.

But caution is called for here. The real world is different from the simulated and rather limited worlds of chess or even go. It is much more difficult to design and build a team of machines that can win the World Cup in soccer than in chess—although it will probably also happen one day. Einstein found the general theory of relativity much easier to master than table tennis. While at Princeton, he asked mathematician Merrill Flood to teach him how to play, but the ball just got tangled in his hair.

In a world full of surprises and unknown rules, you need curiosity to survive. A research team at the Berkeley

Artificial Intelligence Research Lab has developed an algorithm that uses curiosity to learn how to play Super Mario Bros. AlphaZero was rewarded when it won a game of chess, while this software scores if it detects something it cannot predict. It is encouraged to explore new situations. Such a program, however, can get caught staring meaninglessly at a source of unpredictable noise like a flashing TV screen or leaves fluttering in the wind. Too much curiosity can be dangerous if you cannot decide what is relevant and deserves attention. Are the remarkable moves of AlphaZero signs that an alien intelligence has awakened? Probably not. These fantastic programs are no different from my old BASIC program—there are still lines of code—and there is no reason to believe that they contain a consciousness. This is in contrast to the Mechanical Turk, a chess machine that toured the world during the late eighteenth and early nineteenth centuries. Its secret was that a human chess master was hidden within it. But how can we know?

## How Do You Know If Someone Is There?

If you want to determine if a computer has a consciousness, you must have some kind of perception of what a consciousness is. According to Descartes, it is an inner subjective perspective, a "self." When you look at your "self," you have a decidedly privileged perspective.

Instead of immediately becoming entangled in philosophical reasoning, let's tackle the problem as a physicist. If we accept that everything is physics, consciousness, whatever it is, must fit into a physical theory. We must therefore be able to describe how a consciousness relates to other physical phenomena and how it manifests itself in matter. To have any value, the hypothetical theory must also be able to

make new predictions, which will—at least in the future—
be possible to test through measurements.

We begin by pretending that we have no idea that something resembling consciousness exists at all. As uncondi-
tional experimental physicists, we set out in search of new
interesting phenomena without knowing what we might
encounter. We must not engage in anything beyond what
we can observe with our scientific instruments. Our task is
limited to observing, calculating, and setting up models for
what we see. In other words, we are going to devote ourselves
to science. Whether we are studying a rock, a calculator, a
sophisticated computer playing chess, or trying to have an
interesting conversation, we use the same basic models to
describe the material components. Our models show how
the individual parts interact and how what we observe arises.

When we study a stone, we discover the beauty of miner-
alogy. Physics disguised as chemistry explains how hardness
and color can be derived from how the atoms are arranged.
Quantum mechanics plays an important role, and it is certainly not easy to explain the properties of all the different
forms of material that we find in the world. And it is also not
easy to know how to create new ones. Despite this, we are
convinced that the basic fundamental laws of nature have
already been formulated. Similarly, astronomers can model
stars using advanced computer calculations that simulate
nuclear physics in hot, turbulent plasma inside stars. The
stars shine with the help of the same nuclear physics that
melts the Earth's interior. The problems are complex, sometimes approaching the limits of what is known.

A calculator is an invention that is designed and fully
understood by cunning engineers. We need materials physics to understand how to build the components that are
then assembled in a prescribed way. Electric currents flow

through the circuits and perform the mathematical calculations we want them to perform. This also applies to Alpha-Zero, which follows rules that are intentionally written into its software. These enable an evolutionary adaptation, whereby the system gradually improves its ability to play chess. We can marvel at the result and it is possible that the strategies it finds are too complex for us to understand. But the difference between these examples still remains a difference in the degree of complication, not in principle. Everything can be derived from the fundamental laws of physics. There is nothing mysterious about what a rock does or how a computer can beat a person in chess.

What conclusion do we draw about the existence of consciousness? In the description of a rock or a computer playing chess, there is no need for consciousness. We already have everything we need to be able to account for what we observe. Like Democritus, we find that all that exists is atoms and voids.

Let us now take the next step. Presumably, we should be able to apply our argument to biological brains, as well. These also constitute physical systems. The reasonable conclusion should be that a dog does not have an inner life to a greater extent than a computer running a chess program. In fact, this should apply to all living organisms, including humans. All those you love, if the reasoning is correct, are nothing but soulless machines without an inner light. The conclusion is fundamentally terrible but seemingly inevitable.

At this moment, if you turn your gaze inward, you discover something that does not fit among the atoms, the void, and the equations: your own inner subjective presence, your self—the Cartesian phenomenon that you have just ruled out in rocks, computers, and fellow human beings. How this is possible, even though it does not seem to have any place in physics, is what the Australian philosopher David Chalmers calls "the hard problem."

How to resolve the conflict? One possibility is to claim that the problem lies outside of physics and science, and declare that by definition, these are issues that science cannot handle. As a scientist, you must never take the subjective seriously, but must leave this to other disciplines. In accordance with Descartes's instructions, we as researchers must look at the world from the outside. It is a convenient way out that might appeal to those who are religious and allows us to avoid conflict. But for a consistent naturalist, such a dualistic approach is an impossibility. We must seek the solution elsewhere.

Physics is not only about atoms and voids. Physicists use concepts with meaning at higher levels. We do not always have to keep track of individual atoms, or the even more fundamental particles they are made of, when we study the fall of an apple. We refer to the collection of atoms as "an apple" and focus on how the center of mass falls. When we discuss air pressure and temperature, we focus on the behavior of large numbers of molecules. In the same way, a hurricane with its strong spinning winds we understand not in terms of the individual air molecules, but as air volumes that extend over many cubic kilometers. It is a matter of course for every physicist to distinguish phenomena on a larger scale in this way and focus on them. Such phenomena are called "emergent." Simple basic laws governing microscopic constituents can lead to complex processes on macroscopic scales. These, in turn, can be described by new laws that may seem completely unexpected from a microscopic perspective. Can consciousness be considered an emergent phenomenon in the same way?

The philosopher Daniel Dennett is one of those who most

clearly argues for this view. In his *Consciousness Explained* (1991), and in several later books, he presents his arguments. Whatever we experience—call it consciousness if you will—is a consequence of physical processes. Dennett is convinced that physicists like me have already found all the relevant facts we need to understand it. He concludes that it is a simple experimental fact that the existence of a consciousness follows in the presence of sufficiently complicated information processing similar to that which occurs in our brains. An immediate consequence of this is that computers, no matter how they are designed, will also attain consciousness.

The philosopher John Searle has argued against this view, with the help of a famous thought experiment. Suppose there is a computer program that can converse in Chinese. Searle, who does not know a word of Chinese, imagines himself trapped in a room with access to an English version of the program. Using pen and paper, he draws the prescribed manipulations of the Chinese characters that enter through a crack under the door. He then slides the results through the same gap. Searle concludes that the room itself, which includes him, does not understand Chinese or what the conversation is about. Dennett, on the other hand, believes that this is a hasty conclusion, and that Searle describes the experiment in a misleading way. If the room can really carry out the task, it must be extremely sophisticated and, in fact, also conscious, no matter how strange this may seem. Dennett argues that consciousness is a secondary, emergent phenomenon that arises as a result of any sufficiently complex information processing. Dennett presents this astonishing conclusion to finally bring an end to the question of the purpose of consciousness. Science has shown that consciousness is merely an illusion explained by evolution. Period.

Dennett is not alone in putting forward ideas of this kind,

which have even been given the name computational theory of mind, or CTM. To a physicist like myself, the word *theory* does not seem to be accurate. In physics, a theory must be formulated with the help of mathematics and adapted to what we already know. It must, at least in principle, be usable to extract predictions of new phenomena that can be tested with the help of experiments. The general theory of relativity is an example of a theory that lives up to all tests, while string theory is still waiting to be tested.

The very core of CTM derives from the observation that consciousnesses and computations seem to be related in the sense that conscious people can count. It is argued that the relationship is causal and that the mysterious phenomenon of consciousness is caused by the phenomenon of calculations, which we are supposed to understand better. CTM even claims that sufficiently complicated calculations always give rise to a consciousness. But the theory does so without explaining why.

Not even in biological beings is there an obvious connection between consciousness and calculations. Our brain certainly performs activities that can be characterized as calculations without our even being aware of it. And I can feel conscious without engaging in any mathematical operations. The connection is thus far from a given. And even if one could establish a correlation, this is not enough to establish a theory on this basis. The theory gives no testable prediction. Instead of concluding that calculations give rise to consciousness, it seems more reasonable to state that a consciousness in the right mood can perform calculations and that there may also be other ways of calculating that do not involve consciousness at all. That is precisely the province of computers.

What Dennett and others are trying to do is solve the

problem by ignoring it. It is assumed that there is nothing beyond the material world—as a naturalist, I fully agree with that—but to this they add the completely unfounded assumption that we more or less already understand everything about matter through today's physics. From these assumptions, it is then concluded that consciousness arises artificially as soon as computers become sufficiently sophisticated.

The two assumptions may seem innocent to the scientist, but one must completely reject the inner subjective perspective if one is to buy the conclusion. Dennett and others point to the impressive advances in science to provide an authoritarian argument as to why you should not trust your experience. It is an illusion, they claim. But the contradiction is immediate and it is not lessened by the writing of thick books that try to dismiss it. An illusion requires that someone be deceived. An emergent phenomenon presupposes the existence of someone, perhaps a scientist, who identifies the phenomenon and gives it a name. This is not so difficult when it comes to apples and hurricanes, but a completely different thing when it comes to the researcher's own consciousness. It takes one to recognize one. All attempts to explain away the problem make use of exactly the perspective you want to get rid of.

## Are There Zombies?

If consciousness is so debated—some believe that it is something beyond physics (whatever that may mean), while others believe it is an illusion (whatever that may mean)—it is reasonable to ask what it would mean if a person lacked it. David Chalmers discusses the matter in a famous thought experiment where he introduces the concept of a *philosophical zombie*. The word *philosophical* emphasizes that it has nothing to do with the living dead in popular culture. The definition is simple: a being who outwardly acts just like a human but lacks an inner life, a self. This inner world is referred to as "qualia" in the philosophy of consciousness. No matter what experiment you perform, it remains impossible to detect any difference between a zombie and a real person. Would this be possible, even in principle? As we will see, it is not the answer itself that is interesting, but the thoughts the question leads to and how you argue for your position.

Dennett and other supporters of CTM exclude the possibility of zombies based on the reasonable assumption that all traits must leave physical traces. If there is no physical difference—this includes anatomy as well as behavior—between what is assumed to be a zombie and what is assumed to be a real person, there can be no difference at all on any level.

The thoughts, the illusions if you will, must be there as patterns in the zombie's brain as well as in the brain of a human. If one of them is aware, so is the other. To the extent that consciousness can be explained as an illusion, the same applies in both cases. From this point of view, a philosophical zombie is an impossibility.

Others are not as confident and find the explanation problematic. If one cannot deduce the presence of a consciousness with the help of theoretical arguments based on knowledge of known physics, and there is no method to detect it with the help of a measurement, how do I know that it is there? How can I even be sure that those I love are not philosophical zombies?

What other people really think and how they experience the world is not easy to know. One might imagine a correspondence to one's own experiences, but there are some known exceptions. As I have already mentioned, I have a defect that many find difficult to imagine. Like a small percentage of people, I am color-blind. I have a hard time seeing the difference between red and green, especially in poor lighting. This means that the world I see around me, and reasonably the one I can imagine with my inner eye, is not quite as colorful as it is for the majority of people. About 2 percent of all people suffer from another, perhaps stranger condition. It is called aphantasia and involves an inability to create images within oneself, a lack of an "inner eye." Those who have aphantasia are surprised to learn that there are people who can see pictures in their heads. One in fifty of you reading this book has aphantasia. Could you be one of them and this be the first time you have been made aware of it?

Others have an inner eye with greater capability—in essence, a full 4K Ultra HD screen with stereo sound that makes all thoughts and dreams impossible to distinguish from reality. Are you one of the (lucky?) few with this ability? While I'm not, I still have a cinema in my head that plays an important role in everything I do. When I am without pen and paper, whether I am waiting for a train, waking up in the middle of the night, out walking, or simply bored, I can entertain myself by performing mathematical calculations for my inner eye on a little blackboard. I can write and manipulate mathematical equations just as I do on paper, and in some respects it works even better. I can make the symbols dance around and merge with one another with a clarity that is difficult to match with chalk on a real blackboard. I can also add that there is an inner voice that comments on what is happening. Funnily enough, it often uses English instead of Swedish when it comes to mathematical subjects. But there are limitations. As I mentioned, the board is small and the equations tend to fade annoyingly—especially if they are a little more complicated.

Now, let us assume that there is a fundamental difference in the inner life of different people—not only in terms of the ability to imagine colors and images but to what extent there is a consciousness. Imagine that lack of qualia is a fairly common handicap and say that one in a hundred people you meet on the street is affected. Or worse, maybe it's the other way around and you're in the minority. Say one in ten. Or, to take it even further, perhaps in the whole universe there are no more than you and I who have an inner self. How can you be sure that I, who sits and writes this book, am not an empty shell that simulates a consciousness and that you, just you, are the only consciousness that exists at all?

In an influential article from 1950, the inventor of the modern computer, Alan Turing, asked the question "Can computers think?" Or rather, can a computer behave in such a way that it seems able to think? He imagined an imita- tion game, which we now call "the Turing test," in which a computer tries to converse in such a way that it succeeds in convincing a human being that it is another human being and not a computer. If you believe in CTM, it is reasonable to conclude that a computer that passes the Turing test is also aware. Otherwise, you have to accept the possibility of a philosophical zombie, which is contrary to the basic assump- tion. Others, such as John Searle, are not impressed. The test may say something about intelligence, but why would the result have anything to do with consciousness?

Online, you sometimes get the question "Are you a robot?" and you must answer "No" to proceed and be allowed to download material or whatever it is you want to do. There are certainly programs that pass this simple test and it will surely be possible to create programs that not only play chess, go, and poker but can also have a decent conversation. In this context, Google Home and Siri leave quite a lot to be desired. They are like boring know-it-alls who can answer most factual questions but are not good enough for much more. Maybe one day it will be possible to perform a kind of reverse Turing test, where a computer sincerely answers the question "Are you aware?" with a resounding "No!"

## A Third Possibility

We have thus dismissed two popular approaches impossible to reconcile with a naturalistic worldview. One is that consciousness is independent of the physical world; the other is that it is an illusion. How do we move forward? Certainly no new physics is needed to understand how a computer works—it uses existing technology—but biological brains are a completely different matter. We can all make the crucial observation ourselves that the inner subjective perspective actually exists. Descartes also realized this. Therefore, there must be new physics associated with biological systems. This inevitably follows from the assumption that everything is physics and the observation that your consciousness is not an illusion. I can only guess at what this new physics can be. This calls to mind Hempel's dilemma, formulated in 1969 by the German philosopher Carl Hempel (1905–1997).

What does it really mean that everything is physics? A physicalist is a naturalist with a special focus on exploring nature using physical models. As a pure physicist, of course, I want to claim that everything is physics. If physics refers to the physics we know today, the concept becomes time-dependent. If, on the other hand, one includes everything that could possibly be discovered even in the distant future, the concept becomes vague and poorly defined. When Hempel formulated his dilemma, most particle physics belonged to the future, along with the quantum Hall effect or the strange material graphene. In the nineteenth century, people had no idea about quantum mechanics and relativity. Despite this, physics was already a rich subject area, with physicists and philosophers engaged in deep discussions about what it all really meant. If only they could have known what the coming century would bring. Physics continues to evolve and we have no idea what discoveries will be made in the decades or centuries to come. There may be physics that we will never be able to understand. When I define myself as a physicist, this is what I have in mind; I happily live with this dilemma.

I see physics as a theory of everything by definition. To the extent that there is a first-person perspective, physics, as I define it, must include it. The critical question is whether we limited humans will ever master it.

When you look at it this way, the question of zombies can be reconsidered. In line with CTM, I conclude that the physical state reflects everything that exists and that everything is encompassed by physics. There is, therefore, always an opportunity to distinguish a zombie without consciousness from a human with a consciousness through some form of cunning experiment or measurement. Exactly what, we cannot say at present. Both in the zombie's brain and in ours, there must be patterns that we can link to ongoing processes that we would like to identify as thoughts. Brain scans tell a lot about what happens in a brain, and electric currents reveal what calculations take place in a computer. In any case, there must be subtle physical differences between conscious thoughts and pure information processing. Whether we are smart enough to understand where the difference lies is another matter.

In real life, I suspect that the task is much simpler and that even a fairly superficial contact with a zombie would be enough to tell that no one was there. It is reasonable to imagine that there are a lot of tasks that are beyond a zombie, since they require consciousness. For instance, a zombie might have a hard time following discussions about what it is like to have a consciousness. Perhaps conscious beings rather than zombies evolved because they are better able to cope in the real world. Evolution may have discovered how physics offers possibilities that we still have no idea about.

Phenomenologists such as Merleau-Ponty and the German philosopher Max Scheler (1874–1928) do not see the question of how to recognize another consciousness as a particularly serious problem. In fact, most people take the presence of other subjective consciousnesses as a matter of course in everyday life. Their presence is seen as immediate and not as the result of any advanced theoretical analysis. Isn't it a bit backward to question the presence of another consciousness when you still take so much else for granted? Other consciousnesses are nothing but aspects of matter. What you observe when interacting with a friend or playing with your child is a direct effect of the presence of another consciousness. According to Merleau-Ponty, the point is that you are a body, rather than that you have a body. It is only if you are stuck in the deceptive Cartesian dualism that philosophical zombies are a concern that you end up in trouble. Of course, this does not mean that you cannot be deceived into believing that there is a consciousness in a machine, as well. This is exactly the danger that supporters of CTM are exposed to.

## Simulated Worlds

Instead of directly trying to answer the tricky question of whether a computer can think, it is illustrative to divide the question into two parts. First, you sort out the extent to which a computer can simulate a thought, and then you address the question of whether a simulation can be identified with reality itself.

When my eldest son took an interest in World of Warcraft, I decided as a responsible parent to find out what the game was all about. When I was faced with the task of building my own character, I took the opportunity to change gender. I called myself Groa—after a female oracle in Norse mythology—and set out to conquer the world. It was a captivating but scary landscape that spread out in front of me. To avoid being killed, I hid to the best of my ability and never managed to advance beyond level five. Pitiful by all measures.

Others succeed better in simulated worlds. The technology behind computer games continues to evolve, and the illusions are becoming more convincing—and addictive. Some players thrive better in virtual realities than in the real world and take better care of their virtual characters than they do of their physical bodies. It is obvious that it does not take much to deceive the brain into believing that it is in a different place or at a different time than where the physical body is actually located. In the early history of computer games, many believed that equipment such as 3-D glasses would be necessary to create compelling illusions. It soon became clear that the brain could fix much of this on its own. Now, several decades later, technology has matured to the point where it is ready to surpass the imagination and deliver fully credible virtual worlds.

Will it be possible to construct a complete simulation of an entire world, perhaps even our world? I do not mean simplified representations in the form of clumsy figures, odorless and full of pixels, but a reproduction where no details are omitted and it becomes completely impossible to separate the simulation from reality. To achieve this, you need to work your way down to the level of particle physics and even further. A curious particle physicist who builds instruments to scrutinize the simulation would have to be impressed. All data would have to be represented as sequences of numbers that evolve in accordance with algorithms that represent the laws of physics with the required accuracy.

The better the representation, the more it must be self-referential and the more meaningless it becomes. Meaning is created in an exchange with a surrounding physical world, not through computations performed on a computer. Despite this, there are those who want to see the abstract numbers produced by such computations as identical with a living, physical world. Having come this far, it is logical to question whether you even need to run a program on a computer for a simulated world to become real. Is it enough that it can in principle exist for the world it simulates to be real? All sorts of stories, whether told or untold, then, become as real as the one you mistake for reality. Somewhere Frodo throws the ring into the fires of Mordor. Once you have crossed the line between reality and fantasy, there is no reason to stop.

## The Trilemma of Simulation

Just as pure formalism cannot capture the essence of mathematics, modern physics cannot capture how matter produces consciousness through organic beings. Physicists like me have high ambitions and strive for a theory for everything, but at the same time we exclude the inner subjective experience, doubting that it even exists. Gödel proved that mathematics was so much more than manipulating meaningless symbols according to fixed rules. No matter what system you set up, there is always something outside of it. There are truths that cannot be proven but are still true and there are numbers that not even the most powerful computer can calculate. But while mathematicians realized that the dream is impossible, they do not have much interest in the real world. Physicists, on the other hand, ignored what mathematicians discovered and continued to dream.

Physicists tend to fall into two basic traps. One concerns the confusion between models and the world itself. Since the models are formulated with the help of mathematics, it is an understandable mistake to identify the world itself as mathematics. The second trap derives from not fully understanding Gödel's conclusions and keeping Hilbert's hopeless dream of completely mechanical mathematics alive. If the world can be identified with mathematics, the conclusion follows that the world itself is just meaningless syntax, without the need for any semantics.

This produces a number of strange consequences. For example, if everything is just formalism without meaning, there can be no difference between a simulation and the real world. If you take this as a starting point, it is completely rational to worry that we are just a program running on a computer. Movies like *The Matrix* allude to such notions, but this is nothing new. The Argentine Luis Borges (1899–1986) and many other writers have toyed with the idea that the world of fiction is no less real than what we call the real world. While this can result in entertaining and educational thought exercises, it is basically pure nonsense.

The Swedish-born Oxford philosopher Nick Bostrom has, however, argued that we must take seriously the possibility that we live in a simulation. If you imagine a future civilization on Earth that is so advanced that it is capable of simulating entire worlds with all its inhabitants, you might conjecture that such a civilization could embark on simulating its own past, time and time again, changing the parameters to see what happens. This, in turn, would mean that the number of simulated versions of ourselves would be enormous. Therefore, it is more likely that you belong to a simulation than to the real world. Bostrom sums this up in what he calls his "trilemma," where one of three things must be true: Man would be exterminated before this happens, man will never try such a simulation, or, finally, we are actually simulated.

Bostrom is not sure which of the three alternatives is the correct one, but he considers them to be equally probable. For my part, I choose a variant of the second option. Our descendants will never be able to create a simulation that can be confused with reality, for a very simple reason. The calculations that take place in a computer to illustrate the physical phenomena we are interested in are completely different from the phenomena themselves. A simulation depicts reality but can never be identical to it. Consciousness, with its inner subjective experience, is a real physical phenomenon and therefore can never be identical to a simulation.

To further illustrate the unreasonableness of Bostrom's reasoning, let's imagine a computer scientist sitting at her supercomputer after having just implemented an advanced version of a simulation of the world. She carefully and curiously explores the simulation and is struck by how well it reproduces the real world in which she spends her life. Then she has a scary thought and steers the screen's field of view toward a familiar part of the world. She zooms in and manages to find the house where she is currently sitting. She takes a deep breath, increases the magnification even more, and there—or is it here?—she finds herself as she sits and explores the world.

Is this really possible? The real world would thus be a simulation inside a world that is nothing but the world itself—a closed self-sustaining loop. The argument is a perfect parallel to how consciousness is claimed to be an illusion. Consciousness creates an illusion, which is precisely consciousness itself, in the same way that reality simulates itself.

One does not have to discuss consciousnesses to appreciate the difference between reality and simulation. It is enough to look at a single atom. A simulated atom is created by the hardware in a simulator that typically consists of an astronomically large number of atoms. In other words, simulating a single atom requires a scientist who looks at it and, in his imagination, thinks of a complex system as an atom (which it is not).

There are additional ways to understand the importance of keeping reality and simulation separate. It is true that simulated alien worlds of astonishing complexity can be constructed using extremely simple rules. A well-studied example is Game of Life, invented by the British mathematician John Conway. The game can basically be played on a paper grid but works best on a computer. Boxes are filled or left blank according to the rules of the game, and intriguing shapes develop that move across the paper, or the screen, evoking living organisms. Game of Life is a closed universe that is fully determined by the initial conditions it is subject to. Once you have decided which of the grid's boxes are to be filled at the beginning of the game, the history that enfolds when you play the game is determined. The kind of hardware that is used does not matter. So why even run the program? It would seem to be that the simulated world exists, containing whoever lives there, even before you turn on the computer.

That our world is just a simulation inside another world that would be more real is perfectly in line with religious speculations about a supernatural world. Those who defend the possibility from a more scientific point of view can claim that they advocate something entirely naturalistic. Sure, what we experience here and now may be part of a simulated artifact, but this simulation is, in turn, run on a real computer in a real and natural world. What we should ultimately call nature could be something completely different in essence from what we directly experience in our simulated universe. The laws of physics could also be completely different, just as the laws that govern Game of Life have little in common with the standard model of particle physics. This real natural world outside the simulation can be compared with what those who are religious would identify with the supernatural. What, exactly, is the difference?

If you really believe that our world is a simulation, you'd have to admit the possibility that the great programmer has chosen to act like a God, introducing miracles here and there. In this way, belief in our world as a simulation becomes impossible to distinguish from belief in a supernatural God.

This brings to mind a meeting I had with the Oxford mathematician and creationist John Lennox. He is an amiable man with a genuine interest in science and with an extreme creationist worldview. Our discussion was conducted in front of an audience of devout Free Church members and focused on the seemingly fine-tuned natural constants in nature. If, for example, the strength of the electromagnetic force changed only slightly, no stars would shine and no life of the kind we know would be possible in the universe. I argued that the multiverse could be a plausible explanation for this strange fact. If the multiverse were large and varied enough, the laws of nature in some small corner could by chance be hospitable for life. This would be a perfect parallel to the fact that there are many different kinds of planets in our universe. There is no particular reason why the conditions on Earth are so favorable. If there are enough planets, such conditions may well exist somewhere else. John, on the other hand, saw God's hand and believed that there must be an intention behind it. The discussion was, as far as I could judge, rational, respectful, and entertaining—up to a point. When I asked John if there was anything that could make him give up his faith in God, everything took a different turn. After a few seconds, the answer came: if it could be proven that Jesus never rose from the dead. Since this is an alleged historical event, he admitted that it was falsifiable. What such evidence could consist of, on the other hand, was less clear because the authority of the Bible preceded any other conceivable evidence. John's equals have no problem seeing God as the great mathematician who guarantees that our universe is comprehensible. Similarly, I suspect that ideas that our universe is a simulation would feel strangely familiar to Lennox.

## Dangers of the Future

> *I'm afraid. I'm afraid, Dave. Dave, my mind is going. I can feel it. I can feel it. My mind is going. There is no question about it. I can feel it. I can feel it. I can feel it.*
>
> —*2001: A Space Odyssey*

In his novel *Two Years Eight Months and Twenty-Eight Nights*, Salman Rushdie tells a story about how people in a distant and happy future, guided solely by their intellect, rejoice in everyday things. There is only one thing they lack: They no longer dream at night.

IS THERE A RISK THAT OUR WORLD IS HEADING in a similar direction? There are good reasons to discuss the risks associated with the continued development of artificial intelligence. Opinions are divided on what the dangers are. Some speculate that robots will become independent of their human designers and take over the world. Famous researchers such as Stephen Hawking (1942–2018) write appeals to suggest that certain types of research must be banned.

If history has taught us anything, it is that stupidity rather than abundance of intelligence is what leads to disaster. There are more reasons to fear people's reliance on rather boring technical systems with inadequate safety margins than apocalyptic scenarios in which fast computers take command, even if the latter are more inspiring for writers and filmmakers.

Others argue that the robots of the future will act in a more morally competent way than humans. They believe that we should welcome the possibility that if beings without human faults took over, they would make the world a happier place. Why should we nostalgically cling to something hopelessly archaic when it can be replaced by something far better?

I do not share this carefree view of the future and see the belief in such a scenario as the real danger. Through evolution, life has developed a capacity for consciousness that is useful for survival. It is hardly something that is limited to man, but is shared, to varying degrees, by other fellow travelers on planet Earth. In the same way that other qualities and skills have come and gone during the history of the Earth, there is no guarantee that a developed consciousness must necessarily persist.

Nor can I rule out the possibility that humanlike entities with intelligence and intelligent behavior could be realized in ways where consciousness does not play any role. Outwardly, they may seem to be doing exactly what it takes to adapt and survive in the world. Perhaps we would preprogram them with human morality that directs them to act in a way we find exemplary. On the surface, this world may appear perfect, but behind the facade, everything would be just a simulated spectacle.

In Stanley Kubrick's *2001: A Space Odyssey,* the crew member Dave is fighting with the supercomputer HAL 9000. Despite HAL's appeals, Dave finally manages to do the only right thing: He pulls the plug on HAL. In the real world, it might not be that simple. How would I feel in front of a crying robot that borrowed its appearance from a human child? Instinctively, for reasons of survival, we animate our surroundings and assume that others have an inner life just like ours.

How research on artificial intelligence is presented, interpreted, and woven into popular culture has consequences that are existential and affect how we look at ourselves and choose our future. I share to some extent the concerns that exist about what AI can do to us. The big danger, to me, is that we begin to believe that artificially intelligent machines are conscious beings. In the long run, this illusion risks relativizing human values such as human rights. We have, for good reason, begun to recognize the rights of other animals in proportion to their estimated level of consciousness. We have concluded that other mammals are likely to experience pain and suffering in a way similar to ourselves. We are less sure about fish and insects and have no compassion for mowed grass. Nor do I have any for my computer; I try to take good care of it because it was expensive and I have a lot of use for it, but when it becomes obsolete, I won't hesitate to destroy it.

But the more robots resemble humans in the way they behave and interact with us, the more likely we are to be tricked into projecting onto them an inner life similar to our own. We already treat inanimate objects as if they are alive. We curse the car, maybe even hit it when it refuses to start. A child hugs a teddy bear and speaks softly to it. When robots begin to resemble us, will there be political movements that advocate for them to be given the same rights as people, including freedom and protections that go beyond their economic value?

Self-driving cars already raise serious ethical problems. How should they be programmed to act in situations where decisions affect the safety of people in the car versus the safety of pedestrians? Whose fault is it if something goes wrong? Let's imagine a crash with a self-driving car—no technology is perfect—in which a number of people are seriously injured. When the ambulance personnel arrive, they find the car is burning and the injured must quickly be brought to safety to have a chance of survival. In an ordinary accident, no one would bother to save either the car as a whole or the electronic components. But what if the artificial intelligence that controls the car has reached such a level that people have developed a relationship with it? What if the computer in the car has reached a degree of simulated humanity that leads us to consider it conscious? Would it then be reasonable to try to save the computer, even though it increases the risk for others involved? These questions are by no means just harmless philosophical musings. The way in which we view consciousness has far-reaching consequences for how we set up society.

*Eliminative materialism* is a body of thought that claims that our beliefs and our mental state in general have nothing to do with reality. Modern advocates such as the philosophical couple Paul and Patricia Churchland have argued that when we humans interact with one another, we should stop referring to nonexistent subjective states in ourselves; that we're better off scanning the brain and letting an electronic gadget objectively tell us if we are sad or happy. In fact, many of us have already taken steps toward this when we let apps determine whether we are getting enough exercise or if we should increase or decrease our pace during a run.

It is very likely that the ideas we have about who we are, are wrong in many ways. Until half a millennium ago, most people believed that the Sun moved around the Earth, and quantum mechanics was a total surprise when it was discovered a little more than a century ago. Why should we trust our perception of ourselves? The Churchlands advise us to ignore everything introspection tells us. While I agree that much of what we take for granted is illusory, I find it difficult to deny that the existence of subjectivity and consciousness is a fact.

If there is any advice I would give to those who develop systems of artificial intelligence, it would be not to fall for the temptation to make them appear too human. We should be encouraged to recognize them for what they really are: machines without an inner life. Otherwise, we may find ourselves gradually replaced by a form of unconscious intelligence that is in some ways superior to the one that has evolved in the old-fashioned way. One day, will our bodies and brains be so intertwined with new technology that it will be impossible to determine the boundaries between what is animate and what is not?

# 6

# Not Everything Can Be Calculated

*There is a theory which states that if ever anyone
discovers exactly what the Universe is for and why
it is here, it will instantly disappear and be replaced
by something even more bizarre and inexplicable.
There is another theory which states that this has
already happened.*

—Douglas Adams

ABISKO, IN NORTHERN SWEDEN, is one of the few
remaining wildernesses in Europe. I hiked there with
one of my very best friends a few years ago. We have done
many mountain hikes together and it was not our first time
in Abisko. We flew up to Kiruna and then took the train to
the mountain station. We arrived late in the afternoon and
planned to go to a lake high up in a valley located between
two mountains to camp. Immediately after we got off the
train, we put on our backpacks and headed up to the bare
mountain. The forest was sparse and the view became more
and more magnificent. Halfway up the mountain, we realized that we had forgotten the gas for our portable stove.
Going out into the wilderness without the opportunity to
cook hot food was not an option and we had no choice but
to return to buy gas and start again.

Once down at the mountain station, it had gotten late, and we decided to set up the tent and wait until the next day before continuing. The lost day meant that we had to modify our plans. We set off early in the morning in a different direction along a popular hiking trail, which was disappointing for us experienced hikers who had wanted to go straight out into the wilderness. After a few hours of intensely discussing the universe, life, and everything (with apologies to Douglas Adams), we realized that we were on our way straight back to the mountain station. We had, without noticing, left the clearly marked path and walked in an almost complete circle, and found ourselves more or less back at the starting point. Once again, we set out from the beginning with the hope of a better result.

In this enchanting environment—in my memory forever associated with closed loops—a series of scientific seminars were held for a number of years. Just as Abisko is located on the border between civilization and the wilderness, the seminars focused on the borderland of knowledge. I attended a couple of them. Several of the seminars dealt with topics related to complex systems and especially life. There was one person who attended several of the Abisko seminars I would have loved to have met: the theoretical biologist Robert Rosen (1934–1998), who passed away several years before I set foot in Abisko. Robert Rosen wrote what may be the most original and least understood book on fundamental science. It has the suggestive title *Life Itself,* and presents a definition of what a living system really is. The ideas are not simple. Rosen's work has been interpreted and reinterpreted over the years, and one may wonder how many people really understand what he meant or the significance of it. The American mathematician John Casti wrote a review of the book in which he wonders who will read it:

*The overwhelming majority of mainline biologists are likely to find its arguments completely incomprehensible. Even more disheartening, those biologists who do understand the book will probably hate it. . . . Mathematicians will hate it too. . . . And even philosophers are likely to turn up their noses in disdain at Rosen's arguments, since again they run completely counter to almost all conventional wisdoms in the philosophy of biology.*

One who really read the book is the mathematician Anders Karlqvist, former head of the Swedish Polar Research Secretariat and also for many years the scientific adviser to the king of Sweden. Anders realized that Abisko was just the right place for those who want to think new thoughts.

We decided to meet in his home just outside Stockholm. Driving my car from Uppsala, I soon found myself directed by my GPS along a complicated route that took me far out into the countryside. I expected a trip toward the capital but instead had to drive along narrow and winding roads through deep forests. It felt like I was back in Abisko, and I began to worry about driving in circles that would return me to the starting point. My head was full of loops, but eventually the trees dispersed and I was back in civilization.

When I arrived, Anders picked out books and manuscripts linked to Rosen and his research. Some of the material was familiar to me, some new. A view of living organisms was rolled out that was far from the generally accepted one. The argumentation was careful and mathematically precise, with the aim of making a clear distinction between organism and machine. Rosen described how living organisms have an ability to take into account the future, to predict what will happen, and to act accordingly. It was in this that the secret of life was hiding.

## Life Itself

Robert Rosen claimed that we must go all the way back to Aristotle to find our way. For Aristotle, physics was not just about dead things, but about understanding the entire physical world. It is a world that not only contains the planets in the sky, flying arrows, rocks, air, water, and fire. It is a living world, filled with plants, animals, and creatures that wonder about what everything means. Aristotle wanted to describe it all. The book in which he formulated his theory of everything he named *Physica*, meaning "physics."

According to Aristotle, there are four kinds of causes: the material, the effective, the formal, and the teleological. The fourth of these causes is about purpose. It is not easy to translate what Aristotle meant into a modern language, but roughly speaking it is about the following.

The material cause corresponds to the matter that builds up the relevant objects, while the formal cause has to do with their shape. The effective cause corresponds to what we would call causality or causation, and describes how one thing leads to the other. The fourth cause, if there is one, is about the purpose of it all. To illustrate the roles of the various causes, it is perhaps best to use Aristotle's own example of a statue. He notes the material cause to be the bronze that the statue is made of; the formal cause is what the statue is to depict; the effective cause is the sculptor who creates the statue. But this is not all. We must also understand *why* the statue is made. Does the sculptor want to honor a king or someone close to him? Or to be famous or rich? Or maybe he's motivated by the joy of creating.

Aristotle believed that a physicist must study all four causes, but that the fourth is especially important. He sees no fundamental difference between the phenomena he finds in nature and those created by humans. He argues that there must be a purpose to how the different parts of an animal's body interact with one another. Of course, he admits, if the rain destroys the crop, it is a coincidence without purpose. On the other hand, he finds it unreasonable that it would be chance alone that lies behind how living organisms are built. Empedocles' idea—an early precursor to Darwin's natural selection—suggesting that living beings are put together by chance and those that function survive, Aristotle dismisses as unlikely.

In the modern science that followed Descartes, we are satisfied with the first three causes. What is the reason for the apple falling from the branch? The material cause is the matter that makes up the Earth, the apple, and the tree. The formal cause is the form of matter. The effective cause is gravity, which overcomes forces in the apple's stem and causes the apple to fall. For what purpose? None. There are no purposes or intentions in nature; there is no ultimate goal. The apple does not fall so Newton can understand gravity.

Aristotle would find the way we see the world in our day as meager, and he would certainly think that modern physics lacks the power to support the tower of reductionism that extends from the fundamental building blocks all the way up to life and consciousness. Of course, there are phenomena that can be fully described with the help of the material and effective causes—just as a modern physicist would claim. Aristotle is clear on that point. The Moon shines with reflected light from the Sun, and if the Earth blocks the path of light, the Moon darkens. The material cause is the Earth, he explains, and the fact that the Earth gets in the way of the Sun is the effective one. That's the whole story. There is no goal or intention with the lunar eclipse. In the same way, the rain falls, although we can be thankful that it happens so that the crop can grow. It is only the material and effective causes that matter, and we recognize here that Aristotle has the sober attitude of a scientist.

But when it comes to living organisms and how they work, Aristotle is not happy with this explanation. He cannot see how life can reproduce and survive without a fourth cause, a purpose. This has nothing at all to do with any kind of supernatural power, but a physical causal relationship that he believes is required for the world to be comprehensible. The fourth reason is still part of what he calls physics. There are no supernatural beings who rule over the natural world. Everything is physics, but a kind of physics that we still do not understand to this day.

Was Aristotle wrong? There is no physical mystery in how life evolved over millions and billions of years. Evolutionary mechanisms also work in many contexts that have nothing to do with life. In my own work as a theoretical physicist, where I need to solve difficult mathematical equations in string theory, I sometimes have to resort to extraordinary methods. Genetic algorithms look for solutions by allowing themselves to evolve and change. Solutions proposed by the algorithm can mate and create new and better solutions that mutate. The best proposals are allowed to survive, and the calculations converge toward mathematical results that neither I nor the computer had a chance to find using conventional methods.

But what about a living organism in itself, no matter how it originated? Is the living organism just a machine, albeit extremely complicated and produced by evolution? Or is there something fundamental that is missing in our understanding? It is by no means a matter of abandoning the objective and naturalistic science that has developed in recent centuries. There is no need for anything beyond physics to understand either life or consciousness. Aristotle already knew this. The correct question to ask is what physics really is. To move forward, we, like Rosen, must ask ourselves what makes living organisms so special.

## The Ship of Theseus

Living organisms are constantly renewing themselves. Most of the matter we are made of is replaced. While the identity of a machine is carried by the material parts, ultimately the individual atoms, nothing like that can be said about a living organism. An organism is an open system with a constant flow in and out, while a machine is essentially closed.

THE GREEK HISTORIAN PLUTARCH TELLS the story of a ship that brought the great hero Theseus and his company of young Athenians back from Crete. The ship was preserved and exhibited for generations, while the wooden planks were replaced one by one as they rotted and fell apart. In the end, the whole ship had been replaced and the question arose as to whether it was the same ship. Philosophers argued about this for centuries—millennia, even. A slightly more modern equivalent consists of the Vasa ship, King Gustaf II Adolf's pride, which sank during its maiden voyage in 1628. In 1961, it was salvaged from the depths of Stockholm's harbor and is now on display at the Vasa Museum. It is a constant struggle to preserve the ship and prevent it from falling apart. When it was buried in the mud at the bottom of the sea, it could withstand the test of time fairly well. It is a completely different matter when the wood and iron are exposed to fresh air. To prevent the ship from falling apart, four thousand rusty iron bolts have been replaced with new and better ones. The story of Theseus's ship repeats itself.

This time it's about the bolts, but what more will be required over the next few hundred years?

A machine can be repaired and its parts replaced one by one. One can imagine a beloved car where the owner decides to change the engine and in the end there is nothing left of the old car. All this is done intentionally and according to plan. Everything is controlled by independent agents, such as the mechanic in the workshop. The Athenians knew how to cut new wood, while the conservators at the Vasa Museum do their best to forge new bolts that hold the ship together and retain its look.

Living organisms must take care of the renewal and repairs on their own. Small wounds heal and we are constantly dependent on the naturally ongoing renewal of our cells and living tissue. Sometimes we injure ourselves or get sick in a way that our body cannot handle on its own. We go to the dentist to deal with a tooth that hurts, and laser surgery can correct defects in our vision. Surgeons can replace a broken knee or even a broken heart with a mechanical substitute. There will come a day, however, when our bodies can no longer be repaired. This is the inevitable fate of the individual body, even though through our children and their children we can have a form of eternal life.

This characteristic feature of all life was in focus when the Chilean biologists Humberto Maturana (1928–2021) and Francisco Varela (1946–2001) suggested that one can define what life is using the concept of *autopoiesis*. The Greek prefix *auto* means "self" and *poiesis* means "creation." An autopoietic system has the ability to maintain and reproduce itself. An individual has a limited life span, but there is still an unbroken line that can be traced back in time from all the life-forms that exist today to the very first when the Earth was young. The genetic information in the form of

DNA can be copied to a piece of paper or to a computer, while the living cell is irreplaceable.

A machine only needs to be repaired when it breaks down, while the very essence of life is a constant flow of matter in and out that must be maintained. The identity of a living being is not borne by the individual material parts even if it is completely dependent on them. To live is to be in contact with a surrounding world where we are like vortices in flowing water. In the same way, the body of a human being is renewed many times over a lifetime, and individuals are replaced, while the species survives. So far, life on Earth has survived for nearly four billion years.

Can you create machines capable of something similar? A car with an extra accessory that automatically repairs a puncture or a ship that can replace rotting boards on its own? Maybe, but when the mechanism responsible for the repairs breaks down, another mechanism must be available to repair it in turn. And this is ongoing, indefinitely postponing the problem but never really solving it.

In science fiction, so-called von Neumann probes are popular. The worst kind are called "berserkers"—self-replicating spacecraft that travel through the universe with the sole mission of destroying all the life they encounter. Fortunately, we have been spared such disasters in our part of the galaxy and we can only hope that no civilization has been mad enough to create this.

Rosen was inspired by Aristotle and defined living systems as closed with regard to effective causes. This means that nothing that acts on the organism from the outside is needed for it to stay alive. If a machine breaks down, it needs help to be repaired. A mechanic with access to a drawing comes to the rescue and fixes what is broken. The mechanic is the effective cause, or reason why the machine continues to work.

Such a thing is generally not necessary for a self-sufficient living organism. It not only repairs itself but it can also repair the mechanisms it needs to repair itself. Organisms are not closed to material causes. They are open, self-sustaining systems that, like Theseus's ship, are constantly renewed.

What Rosen claimed was that he had succeeded in formulating the concept of *autopoiesis* on a solid mathematical basis. You might think that it is possible to create machines that repair themselves if you were just clever enough. These could then multiply and spread across the planet and perhaps even the entire universe. Rosen's astonishing claim is that it is not possible to create infinitely self-repairing machines within the framework of the physics and technology we now understand and master. No matter how we construct them, they will soon break down, and no matter what self-correcting mechanisms we equip them with, their life span will be limited. Something like the living Earth, which has endured and evolved for four billion years, is not possible given our current ability.

The idea that the world itself could be simulated on an ordinary computer, and all the calculations required to understand nature can be performed on a Turing machine, is usually called "the physical Church-Turing hypothesis." Alonzo Church (1903–1995) was an American mathematician who coformulated this hypothesis with Alan Turing. In other words, all that the universe can do, a good-enough computer can do. Rosen claims that the type of self-sustaining loops needed for a living system cannot be simulated. In a world where the Church-Turing hypothesis applies, there can only be machines but no life. This is the crux of the matter. If Rosen is right, the hypothesis is false and living organisms are the first example of physical systems that violate it. If you want to describe living systems in an exhaustive way, you must therefore use models that extend beyond what any conceivable computer can handle: noncomputable mathematics.

## That Which Cannot Be Calculated

The words of my Princeton supervisor, David Gross, echo in my head: "You must always calculate something." When writing a scientific article in theoretical physics, you need to go beyond words and loose speculation. You must give your ideas a mathematical form. Not to mislead or impress and make those around you believe that you have achieved more than what you have actually done, but as a tool for drawing conclusions that would otherwise have been hidden. Ideally, you should also make numerical predictions that can be compared to experiments. Theoretical physicists like me seldom manage to get that far, and not everything can be calculated. There are problems so difficult that no conceivable computer can solve them in a reasonable amount of time.

David Hilbert not only wanted to formalize mathematics but was also interested in finding the solution to some really difficult problems. In 1900, he presented a list of twenty-three such conundrums, of which the tenth is particularly interesting. It treats the so-called Diophantine equations. Diophantus of Alexandria was a Greek mathematician who lived in the third century A.D. He wrote a book called *Arithmetica* about equations with several variables and with integers as coefficients. The most famous example is probably what is usually called "Fermat's last theorem":

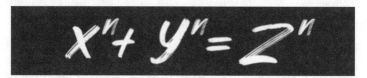

$$x^n + y^n = z^n$$

The task is to find solutions with $n$, $x$, $y$, and $z$ all integers different from zero. If $n = 2$, it is easy to find solutions such as $x = 3$, $y = 4$ and $z = 5$, where you can verify that $3^2 + 4^2 = 9 + 16 = 25 = 5^2$. Strangely enough, nothing like that is possible for integers $n$ greater than 2. The French mathematician Pierre de Fermat (1607–1665) by 1637 concluded that there are in fact no such solutions. He made a note in his copy of *Arithmetica* that "I have a truly wonderful proof of this statement, but the margin is too narrow to accommodate it." No one thinks he really had found a proof, but he has inspired many frustrated mathematicians, amateurs as well as professionals, to search for one. When the British mathematician Andrew Wiles finally succeeded in 1995, with the help of his former student Richard Taylor, it was a bit of an anti-climax and something of a disappointment. Although experts explained that there were extremely original and important mathematics in the proof, it is completely incomprehensible to all but a few.

Here is another example of a Diophantine equation with a long history:

The solution is pretty crazy: $x = 1766319049$ and $y = 226153980$. You have to be in possession of a good calculator even to be able to check the solution. How could anyone have guessed it? Just trying all possibilities out would take an incredibly long time. Amazingly, the first to find it was an Indian mathematician, Jayadeva, in the ninth century. The method he probably used to solve the problem was later named Chakravala by another great mathematician,

Bhaskara, in the twelfth century. Chakravala is the name of a ring of mountains in orbit around the Earth and suggests how respected the mathematical achievement was. *Chakra* is Sanskrit for wheels and refers to a mathematical method used in the proof. Without knowing the background, Fermat failed to find a solution to the problem (this time at least fully aware of his failure). The English mathematician William Brouncker (1620–1684) rediscovered Jayadeva's solution in 1657–1658.

What Hilbert wanted to find was a general method that could determine whether an equation of this kind had an integer solution or not. In the first case, we saw that there was none for integers $n$ greater than 2, while the other had solutions. Unfortunately for Hilbert, no such method exists and every problem is a new challenge.

## A Visit from Douglas Hofstadter

Every Friday at three o'clock, the research group in theoretical physics at Uppsala University has a coffee break, called "fika" in Swedish. According to tradition, someone should present a mathematical or physical problem that the others should solve: *the fika problem*. All conversation shuts down while everyone stares at the whiteboard, at least to begin with. After a while, the discussion picks up speed again when solutions are proposed, tested, and rejected. One Friday, the professor of cognitive science, Douglas Hofstadter, was visiting. Douglas Hofstadter is a hero for anyone interested in math, computers, and logic. His book *Gödel, Escher, Bach* (1979) is a classic and has inspired several generations to study mathematics and computer science. Douglas agreed to contribute a problem and wrote the following innocent equation on the board:

$$\frac{A}{B+C} + \frac{B}{C+A} + \frac{C}{A+B} = 4$$

Find a solution where A, B, and C are positive integers! We studied the equation and began to discuss different ways of manipulating and rewriting it. A suspicion came to me. I asked Douglas, "Do you remember the solution?" The answer came quickly: "No."

We failed to solve the problem for this reason: The smallest numbers that solve the equation are:

A = 154476802108746166441951315019919837485664325669565431700026634898253202035277999

B = 368751317941299998271978115652254748254929799689719709962831374716372246340555 79

C = 437361267792869725786125260237139015281653755816161361862143799337842346777 2036

Insert them into the equation and you will be able to use a computer to verify that it works. These numbers can be compared to the number of protons in the visible universe. What happens if you replace the number 4 on the right-hand side of the equation with other numbers? For odd numbers, there are no solutions at all, while there are solutions for certain even numbers. For the number 178, for example, the solution has 398,605,460 digits, and for the number 896, the solution has more than two trillion. In other words, it would take a couple of million pages of digits to write down the solution to the equation in this particular case. This was by far the most difficult coffee problem ever and it took weeks before everyone had recovered. Diophantine equations are certainly not to be played with, and they lead quickly to mathematics that no ordinary computer can handle.

What about the physics laws themselves? How difficult must the math be to create accurate models of nature? Why would one think it would it be sufficient to limit oneself to the mathematics of the calculable according to the Church-Turing hypothesis? The belief that everything can be calculated on an ordinary computer is somewhat similar to the ancient Greeks' fear of irrational numbers. They did not want to think that there could be numbers, such as *pi,* that could not be written as a fraction. If there is one lesson you can learn from physics, it is that all mathematics will sooner or later be applied. In string theory, the most esoteric number theory emerges if you want to get a grip on the extra dimensions. It is hard to believe that mathematical problems out of reach of Turing machines would forever remain irrelevant to physics.

Maybe one day we'll be able to formulate physical laws whose predictions depend on whether certain hopeless equations have solutions or not. At such time, the theoretical physicist will have difficulty providing the experimentalist with any useful predictions. The mathematics that are relevant to describe some physical systems could go beyond what is calculable even in principle. Meanwhile, the experiment rolls on while nature, seemingly without effort, spits out one result after another and calculates what cannot be calculated.

If Rosen is right, it is enough to look at biological organisms to find examples where this kind of incalculable mathematics that we have discussed is relevant. But an important question remains. How can it be possible for something completely new to sneak into a mechanistic world where everything that happens can be reduced to atoms and voids?

## Everything Big Consists of the Small

*Only entropy comes easy.*
—ANTON CHEKHOV

A few years ago, I did a small tour of Sweden in connection with the publication of one of my books. One evening after giving my talk on the universe, life, and everything, I sat down at a table to sign a few books. I was happy that a fair number of people had lined up to meet me.

I politely asked, as I always do, "To whom should I sign it?" An enthusiastic young woman held out her copy of the book and said, "Write a truth!" Surprised by her unusual request, I hesitated a little. After a few seconds, I wrote:

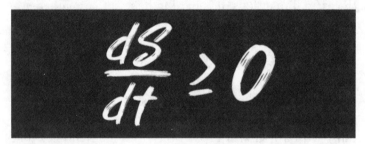

The equation states that entropy can only increase over time. Entropy is a measure of disorder. The greater the entropy, the messier it is. In other words, nothing improves with time.

This is the most basic of all the laws of nature and is called the second law of thermodynamics. The first law says that energy can never be created or destroyed, only transformed. At the end of the nineteenth century, physicists such as Ernst Mach (1838–1916) argued that the laws of thermodynamics were fundamental and could not be reduced to, or explained by, anything more fundamental. There is something appealing and bittersweet in the universality and relentlessness of these laws. On the other hand, these physicists were very skeptical about the existence of microscopic constituents such as atoms.

Despite this skepticism, the Austrian physicist Ludwig Boltzmann (1844–1906) succeeded in showing that the whole of thermodynamics was a simple consequence of Newton's laws applied to a large number of atoms. If you want to keep track of every atom in a gas, it becomes increasingly difficult the more atoms you deal with. In reality, of course, one does not care so much about what the individual atoms do and sums up the behavior of the gas in general with the help of quantities such as pressure and temperature. What Boltzmann showed was that pressure and temperature, and the laws that govern them, are not fundamental in themselves but can be reduced to a microscopic world of atoms in constant motion. The second law simply follows from the fact that everything big consists of small things and that every random change is generally for the worse. Nothing repairs itself, but everything breaks down on its own.

One might think that the second law runs counter to the existence of organic life. Isn't life precisely about creating order out of chaos? Soil is transformed into orderly structures in the form of beautiful and ingenious living organisms. A growing child, the creation of civilizations, and the fine arts must be order's victory over disorder. How can this be reconciled with the second theorem of thermodynamics?

On the whole, disorder always increases, but in small oases, order may temporarily increase at the expense of more disorder elsewhere. The Earth is just such an oasis.

Order is added through the high-quality and well-ordered sunlight where there are few photons of high energy, and disorder is thrown back out into space by heat radiation where there are many photons of low energy. Photosynthesizing plants do their thing, enabling life on Earth to flourish. In a completely closed system, life cannot survive. If the Sun stopped shining, even if we found a way to keep warm, we would still die. Time goes from order to disorder, and in this way a direction in time is created. We remember the past and try to predict the future.

Although we understand how the second law of thermodynamics can be derived from the world of atoms, it has nevertheless retained its special position. No matter how physics will develop in the future, the second law will continue to play a decisive role.

## Emergency—Strong or Weak

The second law of thermodynamics is remarkable. Because it does not work on the small scale of individual particles but is applicable only to large systems consisting of many particles, it describes many phenomena that we would not be able to imagine otherwise.

A SCHOOL OF FISH OR A FLOCK OF BIRDS consists of hundreds or thousands of individuals that move in a seemingly coordinated way. The collective movement is emergent and can be explained by how the individual fish or bird changes its own movement in response to what its nearest neighbors do. Simple models characterized by just a few parameters that describe how an individual acts are sufficient to reproduce these impressive phenomena. The movement of a school of fish depends on how the individual fish adapt their distance and relative angle to other fish. Just a small change in how they swim can turn a chaotic swarm into a circular motion or make all the fish swim away in the same direction.

Can all macroscopic phenomena be explained in the same way and be possible to reduce to the action of simple laws on microscopic scales? The way a living organism is organized and behaves would then in principle not differ from a shoal of fish. Everything we observe could be explained by how a large number of small machines interact, consistent with the belief that consciousness also arises in just such a way.

Most scientists, including physicists, accept that qualitatively new phenomena, and also what we can call new laws of nature, can arise at higher levels. The tricky thing is to distinguish between what is usually called weak and strong emergence. Weak emergence means that the processes at the higher levels obey laws that can be derived from other, more fundamental laws, typically at the microscopic level. The new laws do not really add anything new that does not follow from the fundamental laws. The second law of thermodynamics is just such a law. Some would argue that phenomena such as life and consciousness also belong to weakly emerging phenomena. Strong emergence—or better, ontological emergence—is another and more serious matter. In this case, the laws you are dealing with cannot be derived from any lower level. You find new phenomena that are added at the higher level, not reflected in, or dependent on, the lower ones.

According to causal emergence, there is a ladder of scales where phenomena on one scale depend on what happens on a scale below it. Each scale has a causal structure that makes sense on its own. This allows you to formulate useful laws to describe physical phenomena. The American neuroscientist Erik Hoel argues that the causal structure at one level is essentially independent of that at the lower levels. Biology and psychology have their laws and float on top of physics, unconcerned with the microscopic details. The higher levels contain more information than the lower levels and there are causal structures that cannot be captured by microphysics. Hoel finds support for this through precise mathematical arguments.

Even the most extreme reductionist, or metaphysical realist, will likely agree that the higher steps on the ladder are of great practical value. They provide better ways to keep track of what belongs to everyday life, and the mathematical modeling is much more efficient. The reductionist will be in a hurry to add that in reality there is nothing but atoms and voids and that nothing really new is added at the higher levels, and while it may be wise to use the emerging laws, basically everything is taken care of by particle physics. But is this really true? Could it be that there is physics at the higher levels that cannot be reduced, even in principle? Not only because it is difficult in practice to keep track of what is happening at the lower levels but also because the information is simply not there?

The philosopher Jaegwon Kim (1934–2019) pointed to a fundamental problem that supporters of strong emergence must face. His simple argument is about compatibility. For something to happen at a higher level, something must also happen at a lower level. If the lower level is already unambiguously determined by its own laws of cause and effect, then everything that happens at the higher levels are enslaved epiphenomena. The contradiction, therefore, is profound between strong emergence and the Newtonian paradigm.

Is there, then, no possibility for the higher level to have a retroactive effect? If you could relax the laws a little bit at the bottom of the world, maybe there would be a chance that strong emergence could work. If the atoms in your hand move in response to complex phenomena on a larger scale, the laws governing individual atoms cannot be fully determined by the microscopic level.

American neuroanthropologist Terrence Deacon touches on similar ideas in his book *Incomplete Nature*. He uses quantum mechanics to show that the arguments against strong emergence are probably not so strong. He has a point, but there is another path, independent of quantum mechanics, that has to do with open and closed systems.

## Open or Closed

In a completely closed system, nothing comes in and nothing comes out. Matter and energy are trapped, and so is information. A closed system does not depend on anything outside of it and nothing that happens inside it affects anything outside. Such systems do not exist in reality. The smaller the hypothetical system, and the shorter the period of time involved, the easier it is to isolate the system. As soon as we observe the world, we face an uncontrollable connection with the surrounding universe.

The way we translate between a scientific model and the real world is not trivial at all, but something that is rarely discussed and often actively ignored. Instead, we tend to take for granted that our mathematical theories can be identified with the world itself. Not only is it seen to be practically irrelevant to maintain the distinction; the claim is that the identification of the model with what actually exists says something profound about the world. Mathematics is seen as the language of nature, and nothing is assumed to exist if it cannot be expressed within the framework of the mathematical model. Some go so far as to equate the mathematical representation with the real world and say that everything is, in fact, mathematics. They argue that the formal and natural systems cannot be separated, simply because they are identical.

As a physicist, I want to observe and create models. It is not just about particles, and perhaps strings, deep down in the microscopic world, but also about emerging structures at a higher level that are suitable for mathematical modeling. Thermodynamics is a favorite; I fully understand how it follows that a large number of particles affect one another. At even higher levels, where I live my daily life, eat, sleep, walk, and play with my children, I use even more coarse-grained models, even though they may not be very scientific. Living organisms, such as dogs or other humans, are important conglomerates of myriads of particles that I conceptualize as individual objects. And I think of my consciousness, full of thoughts and desires, similarly.

If the system is open, in constant and unpredictable interaction with the universe, the claim that everything in principle can be derived from the microscopic level is an empty one. The statement can never be tested. An organism that would be sufficiently isolated to be measured and controlled could only be dead. A physically meaningful description of a living being is fundamentally different from that of a machine. A complex organic system is in every practical sense strongly emergent.

The crucial point is that there is a difference between models and reality. We are in the middle of a world, which we can never escape, and we can only try to learn and understand as much as possible with our biologically limited abilities. I may be a physicist, but I do not think we know the physics required for us to fully understand the universe. And I'm not sure we ever will.

# 7

## Man Is Not Unique

*It is in moments of illness that we are compelled to recognise that we live not alone but chained to a creature of a different kingdom, whole worlds apart, who has no knowledge of us and by whom it is impossible to make ourselves understood: our body.*

—*Remembrance of Things Past: The Guermantes Way,*
Marcel Proust (tr. C. K. Scott Moncrieff)

When I get up in the morning, I am in the habit of taking a close look at myself in the mirror. It's a necessary safety precaution when I shave, and when I brush my teeth, I have nothing better to do. Usually I don't think twice about it, but occasionally something strange and unpleasant occurs. While I stand there looking at myself, the world seems to stop and I experience a feeling of the surreal. I can only describe it as a rare, brutal confrontation between the inner and the outer image of the world, the first- and third-person perspective. When we look at other people, we see them from the outside and project inner life into them and we are convinced that they are looking back at us. We see ourselves from a completely different point of view, that of our inner self pondering its existence as it pays attention to its own thoughts. Mirrors have a strange, almost magical ability to let the two perspectives collide.

Even more destabilizing is the endless line of repeated images that disappear into the distance like a long corridor as you hold up a mirror in front of another mirror to see yourself looking at yourself looking at yourself.

I'm not sure if Descartes ever used a mirror in his attempts at introspection—mirrors were not as common in his day. In any case, he missed several important points that have confused the discussions about consciousness ever since. Of his conclusions, there is only one I buy, although it is important: the obvious existence of a subjective first-person perspective. I do not find his other assertions as convincing. Descartes argued that his consciousness was independent of his body and not at all identical to it. He based the conclusion on his ability to imagine a subjective consciousness outside the body or even how it manifested itself inside someone else's. I do not agree that something has to be true just because I can imagine it. Thinking about how to fly by flapping your arms does not make it possible to fly. Yet his conclusion survives even now, when we dream of freeing our consciousnesses from our organic bodies and reinserting them into a more durable form.

Instead of looking for what separates us humans from other beings on Earth, it may be more enlightening to focus on what we have in common. By studying others, we deepen our understanding of ourselves and gain important information as to who we are. It is clear that our biological nature is central to our view of the universe. Our consciousness is in our bodies, and the world we experience through our senses is created by our using organic systems that have evolved over millions of years. We are part of a living continuum that stretches back to the very simplest organisms.

All of this is crucial to our understanding of the physical world—the only world that exists.

The ability of animals to get to know and understand the world is determined by how their brains, senses, and bodies are structured. In the words of the American philosopher Thomas Nagel, we can never really understand what it's like to be a bat. None of the experiments we can perform can directly detect the inner experience of someone else. All conclusions must be indirect. We can ask our fellow human beings how they feel and, based on their answers and our shared experiences and biological origins, we can make well-founded guesses about what it is like to be another human—though we can never be certain we fully understand.

The British primatologist Jane Goodall lived for many years with chimpanzees in Tanzania's Gombe Stream National Park. She discovered how individual chimpanzies had distinct personalities, just like humans, how they cared for one another and experienced happiness or sorrow. They could also make tools and, like humans, be brutal and warlike.

Goodall has been criticized for projecting human emotions onto nonhuman animals and for distorting the scientific method by naming chimpanzees instead of merely numbering them. There is a paradox in this, when the distant third-person perspective of science gets in the way of an objective study of the subjective worlds of other beings. What Goodall did was test the scientific hypothesis that our inner worlds do not differ so much from those of the chimpanzees. There is a continuity that includes matter, life, and consciousness, where everything consists of different aspects of the same thing. There are mirrors if you know where to look.

## To Think Like an Octopus

When a new Belgian restaurant opened in town, special-
izing in the typical dish *moules frites,* a personal favorite,
my family went there to try it. The mussels were excellent,
almost as good as the fries, but after eating a few mussels,
I got an unpleasant feeling. There was nothing wrong with
the mussels, but I began to ponder a somewhat disturbing
fact. Mussels have eyes; they are animals that can see. When
preparing mussels at home, you need to make sure they are
alive when you cook them. If the shell is closed, there is a
good chance that the mussel is alive. If the shell is open, tap
it to see if it closes; if not, it is dead, and it is best to throw
it away. Once you have cooked the mussels, keep only those
whose shells have opened. You can be sure that these were
alive and well if their shells opened before the water started
to boil—possibly under great torment. Although the cruel
reality behind *moules frites* was familiar to me, it was suddenly a thought that made me uncomfortable. Could the
mussels have seen a hand approaching the shell to tap it?

I am convinced that it makes sense to ask "What is it
like to be a mussel?" even if the answer is impossible to confirm. When I try, I imagine a foggy and indistinct presence,
reminiscent of the feeling of waking up without knowing
where you are. One thing is for sure: Mussels are not great
thinkers, and the step from them to us is great.

When we trace our own origins back in time, the size
of our brains decreases and, probably, our intelligence. The
journey takes us back more than half a billion years before
we reach our and the mussels' common ancestor. We have to
climb down through the tree of life almost all the way to the
root to climb up again along another of its great branches.
Soon we pass the mussels, which I ate with such reluctance,
to find, a little higher up, the octopuses.

If you want to meet an alien intelligence, you should get to know an octopus. They are comparable in intelligence to cats, although their brains have evolved completely independently of our human brains. Our common origin did not have much to offer in terms of thinking ability, and evolution found two completely different ways to solve the problem of how to make matter think.

What is it like to be an octopus? We can try to guess. The nervous system of an octopus is much more spread out than ours. While we mainly think with the brain, an octopus thinks with its entire body. Sometimes we rely on our reflexes, which aren't exactly intelligence, while they are a response to input, and involve information processing and output. The arms of an octopus can do much more. A better comparison might be walking, or even writing. Once you have learned to walk, it is largely automatic. Your brain is still involved, and you use your eyes to decide where to put your feet, but you don't need to engage much of your conscious attention. When I write these lines, I concentrate on what it is I should write and not how I should move my fingers to hit the right keys. Similarly, but to a much greater extent, the octopus's arms can independently perform all sorts of tasks without the octopus involving its central brain. In addition, an octopus not only thinks with its body; it also uses its body to express its innermost thoughts in a colorful way. (Although octopuses have long been thought to be color-blind and it is not known exactly how or if they can interpret these complex signals.)

There is not one correct way to look at reality. How we systematize and shape concepts about what is objective depends on who we are. The German biologist Jakob von Uexküll (1864–1944) believed that each organism has its species-specific world of phenomena, an *Umwelt*.

Depending on an organism's senses and the environment in which it lives, it carves out its own representation of the world that allows it to survive. As human beings, we have certain things in common that enable us to communicate. Not least, we agree on what we consider to be truly existing objects. Some may try to deny that there really are such things as chairs, books, and people. They argue that everything consists only of atoms and voids, quarks or strings, or what are currently considered to be the fundamental building blocks. Everything else is the result of arbitrary constructions and conventions. But how these hard-line reductionists and metaphysical realists think and feel in everyday life is still based on a worldview that is biologically based in their brains and bodies. Of course, some things depend on culture and experience, but much of our *Umwelt* is common to us all and is inherent in our biological bodies, senses, and brains. Other animals, whether octopuses, mussels, or bats, have a disposition that is completely different in design. All their models are realistic images of the world, true in their own way, and necessary for their survival.

Although such inner worlds may seem foreign to us, there is another but more foreign form of intelligence that is so different that we hardly even recognize it. Plants are physical systems that process a large amount of information in order to survive and grow. Root tips find their way through the soil to maximize the uptake of nutrients, and the fused networks in the ground under a forest have a complexity that surpasses the brain of an animal. Try to imagine an answer to the question "What is it like to be a forest?"

## Mathematics in the Flesh

Our physical bodies determine how we think, what we experience, and who we are. For some, it may be an unpleasant thought that there are other beings that can also think and feel but have inner worlds fundamentally different from ours. An important ingredient in the inner world for those like me is mathematics. Because mathematics and physics, or rather how we perceive them, are dependent on our human nature, we can say that biology comes before physics and mathematics. The way we understand mathematics is embodied: Even the most sophisticated mathematical concepts can be broken down and reduced to simple components, all of which refer to bodily experiences.

One day when I was about to leave my youngest son at preschool, one of the teachers stopped me to ask how to think about infinity. For her, a universe without end was unimaginable. I suggested that the best way to approach the issue was from a human perspective, to think of reaching farther and farther away, without ever achieving contact with the unattainable infinity, but while this abstract mathematical concept may be interesting in itself, it has very little to do with what a practicing physicist does. In the real world, we never handle real infinity. Everything has an end, even if it is not yet in sight. She looked interested, though her gaze became vague. "Someone told me about a type of African music that has no end." I thought about it. Music that never ends was a way to anchor the concept of infinity in practical experience. "Well, sooner or later," I said, "the musicians will get tired of playing. Even if you don't know when. It's the same with the universe. It can't keep going forever." I think she got the point.

Our understanding of infinity goes all the way back to Aristotle, who spoke of potential infinities as processes that can last for any length of time. Counting is an example of this. You can count 1, 2, 3, 4 and continue for as long as you want. When you quit, there is still always the possibility to count one more. Real, actual infinities are another matter. They are just imaginary end points of processes that are repeated time and time again but will never end in our world. When you think of infinity in this way, you can easily understand it and conceptualize it with your body.

We are explorers who are trapped in our organic bodies—or rather, who are nothing but our organic bodies. What "we" are is defined by the same kind of matter that we try to explore. We can never step out of our bodies, let our senses float freely, and see the world from the outside. The philosopher Daniel Dennett, who believes that we cannot trust our inner experiences, is right in a sense. Our experience of an inner subjective self is real and true, but beyond that we are all affected by a stubborn illusion, that of not being tied to a body.

## I Move, I Talk

We have already touched on how the Turing test presupposes a simple relationship between consciousness and the ability to perform various intellectual tasks. When it comes to raw calculating power, we have long ago lost games like chess to computers. It may be more relevant to focus on language and the ability to express thoughts and feelings. If computers could succeed in conducting sophisticated conversations, it would be presumed that they are at least as smart, and conscious, as a university professor.

But the problem of consciousness seems to be just as difficult to solve whether the subject has a university degree or not, and we get no help from intelligence tests. Consciousness, with or without the presence of a self, is not a question of yes or no. The philosopher Maxine Sheets-Johnstone explores the possibility of linguistic thought processes in her book *The Primacy of Movement*. Long before even a rudimentary ability to use language has developed, there may be a perception of an I or a self. I have no difficulty in imagining such a self, where I have lost all traces of a language but am still conscious. What defines one is not the ability to speak or to think, but to move. The ability to move is the basic ability humans share with all other living beings, even if not all of them have a recognizable "self." Further down the steps that lead to human consciousness and beyond, there is a point where *I* in *I move* is lost and only the act of moving remains, together with a consciousness that may not even recognize an *I*. Sometimes when I have been out running, I have no idea where I have been or how I got there and home again. Running, including keeping myself on the road, avoiding running into someone else, or being run over by a car as I cross the street, is largely done on autopilot, except for a few brief elements of conscious action. The *I* in *I run* has given way to the concept of running itself.

Instead of assuming that the presence of self is solely due to intelligence, including mathematical or linguistic ability, a more fruitful starting point for a discussion of consciousness may be movement. Moving is no less an achievement than talking. The first thing you need for movement is a body. When I awaken, I'm usually a little disoriented on my return from the dream world. My head and feet feel far apart. It usually takes a few seconds to focus and figure out where I am. Sometimes I deliberately try to delay the process and imagine that I am lying in one of the many other beds where I have spent nights in my life. Rooms of different sizes, with different lighting, different smells, with a wall on one side or the other. My body will feel smaller if I try to imagine a childhood bed.

The fact that we are all physical bodies can sometimes lead to sudden and striking insights. The French existentialist Jean-Paul Sartre (1905–1980) writes in the novel *Nausea*:

> *I see my hand spread out on the table. It is alive—*
> *it is me. It opens, the fingers unfold and point.*
> *It is lying on its back. It shows me its fat under-*
> *belly. It looks like an animal upside down. The*
> *fingers are the paws. I amuse myself by making*
> *them move about very quickly, like the claws of*
> *a crab which has fallen on its back. . . . I feel my*
> *hand. It is me, those two animals moving about*
> *at the end of my arms.* (tr. Robert Baldick)

A hand is a thing with a strange double nature. It is a material object no different from any other physical object and fully described with the help of the laws of nature. But unlike a book, or any other object separate from the body, the hand can be given movement and be controlled by our thoughts. Without any real effort at all, I can awaken my hand, which is next to the book, and have it pick up the otherwise immobile book. The Dutch artist Maurits Cornelis Escher (1898–1972) describes precisely this relationship when he depicts his own drawing hand in 1948. The hand produces its own physical existence by drawing itself. It is precisely this self-referential and self-sustaining ability that defines living matter.

The German philosopher Martin Heidegger (1889–1976) takes these insights one step further when he points to the contrast between Dasein, or "being," with the Cartesian subject separated from the rest of the world. Dasein, which is characterized by existing in the world and also being mine, is interwoven with the material world and does not have to end at my fingertips. Heidegger tells a story about a carpenter who strikes a nail with a hammer. The hammer becomes an extension of his hand and his body when the movement of the hand is transferred to the hammer. If the carpenter starts thinking of the hammer as a thing, he hits his thumb.

One can never understand consciousness as isolated from the body or the environment. This is illustrated by the difficulty of understanding what it would be like to be a bat. If we ever find a way to create an artificial consciousness similar to that of a human being, it will have to exist along with a body similar to ours.

## A Brain in a Bowl

How can you be sure that you are not a brain in a bowl connected to equipment that simulates everything you experience? One of the first to discuss this terrible possibility was Hilary Putnam in 1981. He argued strongly against the possibility by claiming that if you were really a brain in a bowl—that never knew of a different existence—then you would not be able to understand either what a brain or a bowl is. If concepts cannot exist for themselves, without some form of connection to something that actually exists, then the brain in the bowl becomes a contradiction in terms.

It would seem that Daniel Dennett is not so sure about this. In his entertaining and rather frightening science-fiction short story "Where Am I?" Dennett describes an isolated, living brain in a bowl in a laboratory, a brain that can communicate with and influence the surrounding outside world. The leader of the experiment gives the brain remote control, along with the illusion of being in a place other than the bowl, suggesting that we, too, may be brains in bowls guarded by slimy aliens.

Thought experiments of this kind can be useful but are sometimes misleading. Even if we are unnerved by them, it is important to examine the assumed rules of the game. This is important in physics, where one often solves alleged contradictions through the realization that it is the thought experiment itself that is meaningless.

What is needed for the reasoning to hold is that the brain be connected in such a way that the bowl perfectly simulates, from the brain's perspective, the presence of a body. This includes not only nerve signals in both directions but also chemical substances in the blood, which is another way for the brain to keep in touch with the rest of the body to make decisions. What is required, then, is to completely crack the code for how the brain communicates, and create an artificial body.

There is, however, an easier way to accomplish this. Keep the original biological body with its interface with the physical world and try to fool it. You could, for instance, provide the brain, together with its body, a good book to sink into. This is a cheap and effective way to imagine that you are somewhere else or are someone else. Alternatively, you can use less subtle methods such as TV, computer games, 3-D glasses or perhaps pressure-sensitive and motorized full-coverage suits. Such more realistic versions of the experiment with the brain in the bowl can definitely lead to illusions that make the question "Where am I?" truly interesting.

As the technology of creating virtual worlds evolves, we will increasingly have reason to ask ourselves that question. We are easy to fool. When embedded in the fictional worlds of computer games, the real world can lose some of its charm and feel less real and urgent. With 3-D glasses, you can transport yourself to alien worlds and transform yourself into other kinds of creatures. Anyone who believes in CTM, the notion that the brain works only with calculations, finds it difficult to acknowledge that consciousness is located in the brain. But if consciousness is an illusion, might not the illusion be somewhere else? Who can claim with certainty that we are wrong?

## Continuity of Life

All life must prepare for changes in its environment in order to survive in the long run. It must create models that can be used to predict what may happen in the parts of the environment that affects its survival. This applies to beings like ourselves, where our cognitive abilities, sometimes in the form of mathematics, can be used for activities such as science. This also applies to other life-forms, such as swallows that catch flies or flowers that close when darkness falls. Strategies for survival are built into our hardware as a result of evolution. The growing child needs to adjust its internal parameters to learn to walk steadily. And certain models must be reformulated, over and over again, by each new generation. Our civilization trusts the ability to transfer knowledge orally, in writing and through computers. In the end, it's all about the same thing: the anticipation of change.

Various life-forms focus differently. One can imagine aliens, with ways of thinking different from our own, modeling and predicting a future far beyond ours. But rather than trying to define what is exclusively human and what makes us unique compared to other species, there is more to learn, I believe, by identifying what we have in common with those other species—even bats.

# 8

## Does Free Will Exist?

*Do not worry. Even if you were weighed down by a soul, it desires nothing more than a deep, dreamless sleep. The unloved body will no longer feel any pain. But flesh, skeleton and skin, everything will turn to ashes and also the brain will finally stop thinking. That is why we thank God, who does not exist. Do not worry—everything is in vain, as for all people before you. An ordinary story.*
— MARLEN HAUSHOFER IN *Austrian Monthly*, 47/48 (JULY/AUGUST 1970) (TR. ULF DANIELSSON)

LET'S DO AN EXPERIMENT. Ready? If you believe in free will, raise your right hand. You have two seconds to decide.

Whatever you believe, I am sure you have reasons. They may go back to your childhood, or, on another level, involve neural processes in your brain, or even individual atoms. When your hand moves, there is a lot that can be described with the help of simple physics and chemistry. Depending on the force exerted by the muscles, you can calculate the acceleration and speed of your hand as it moves upward. You can describe how the muscle itself works, how the cells contract and relax in response to electrical and chemical signals.

The problem of free will has less to do with how, or if, we can locate its origins than with how it is defined. What do you mean by free will? When I am asked why I made a certain decision, or acted in a certain way—especially when it comes to something that is important to me—I can usually state my reasons. These can be purely rational. I eat because I'm hungry; I do the dishes because I like a clean kitchen or to keep someone from getting angry. The reasons I read *The Magic Mountain,* by Thomas Mann, every fifteen years are more complex, a result of my experiences through life and how my brain is structured. That, and when I was a teenager, the father of a good friend planted an idea in my mind that it was a book that should be periodically reread.

Does the fact that I can explain my actions, at least to some extent, reduce the importance of free will?

You can rationalize your actions in many ways, but the fact remains that you could have acted differently. Or could you have? You chose to do what you chose to do and you can never repeat the experiment under exactly the same conditions. As a physicist, I know how important it is not to be careless with reasoning based on "What if?"

At the heart of the problem is the conflict between our intuitive sense of freedom in how we act and the seemingly deterministic laws of nature that affect what choices we make. If everything is just a consequence of the laws of physics, it is difficult to see how an action can be free. In order for free will in the ordinary sense to exist, one must imagine an entity— we can call it a soul—that acts outside the laws of nature and is not governed by them. Defined this way, free will is immediately linked to Cartesian dualism. The soul, with its free will, acts outside the laws of nature in much the same way that God intervenes with divine miracles.

The way out, if you insist on free will, must be some kind of gap in the deterministic laws that leaves a certain space where what will happen is not unambiguously determined. Quantum mechanics has an element of absolute chance that could possibly be useful. Instead of leaving everything to this absolute chance, free will could intervene and deliberately control chance at the microscopic level in order to achieve an intended result. No matter how one thinks that free will can be made compatible with the known physical world, the concept of free will assumes that it is something that by definition exists outside the physical world.

A deterministic world seems at first glance more reasonable from a naturalistic perspective. However, it is really not that simple. The basic problem is the confusion of model and reality that it entails. A model can definitely be deterministic, but whether reality itself is fully deterministic is something that in practice is unverifiable. Deterministic models can be tested and found to be valid within limited conditions but not more than that. If one claims that these practical limitations are insignificant, and that it is sufficient to refer to what applies in principle, one lets the dualistic ghost back in. One imagines a demon, a kind of deputy observer, with access to all conceivable measurement data, who with the help of unlimited computational ability can predict the determined future. We have thus left naturalism. Our reference to "in principle" means that we state that something fundamentally untestable must be true. This, like a religious belief, is not compatible with a scientific approach. The demon is nothing more than a supernatural and extraterrestrial entity that we must resort to in order to grasp determinism. Every physicist knows that this is an unlawful violation of the rules of the game.

The determinist claims that everything, including the

choices one makes, is in principle determined because one can imagine a being who knows everything. The one who defends free will, on the other hand, claims that one could in principle have made a different choice than one did. But whether or not one could have chosen differently is beside the point.

It is paradoxical that both free will and determinism depend on the dualism against which this book is a single long argument. In order for free will to really be free or determinism completely determined and unfree, a universal validity is required. From the perspective for which I argue, the two views are equally naïve and impossible, because both are based on an unattainable omniscient perspective and an outdated dualism.

The concept of free will is based on a worldview where the boundary between models and the world itself is not clear. We draw conclusions about the world based on models that will never be complete. Even when we acknowledge our limited knowledge, we fall into language and ways of thinking that are inadequate to the task. The physicist who wants to test the deterministic laws relies on verification through experimentation. When you perform an experiment, you set up certain initial conditions; then you run the experiment and register what happens. After that, you may change some parameters, run the experiment again, and note the result. What enables you to do this may appear to be your free will, which sits in the soul that Descartes described. It is indeed ironic how these supposedly essentially different concepts merge here.

Yet we cannot extract ourselves from the world, but must remain in the middle of it. Whatever happens, we play our roles and try to understand our experience. We, our consciousnesses included, are an important part of the world itself. We are not slaves to the laws of nature. The laws of nature are just our way of describing what nature does, including ourselves. The naturalist stands in the middle of the world, in the middle of the universe, and tries as best he can, trapped in his mortal body but with the help of an imperfect brain, to articulate what he observes. All models have limitations, which, when reached, lead to new questions.

The apple does not fall because the law of gravity forces it to do so. The apple falls and the law of gravity describes what we observe. It is the same with all other physical phenomena. When you want something and make a decision, it may in some cases be possible to derive your choice from factors in your environment or in yourself and your biological and psychological nature. An attentive observer may be able to predict your future actions successfully. We constantly erect more or less successful models to predict the behaviors of our fellow human beings. But they will do what they will do, regardless of our models. Just as the apple falls, the Earth moves around the Sun, and the universe expands, you make your choices.

There is a story of somewhat unclear origin about a scorpion carried across a river by a frog. There is an older version that goes back at least fifteen hundred years that features a turtle instead of a frog. The frog, or turtle, worries that the scorpion will sting it and that they will both drown. The scorpion assures the frog that it would not do anything so stupid, but once out on the river, it stings the frog. When they are both about to perish in the waves, the frog asks the scorpion why he did this. The scorpion replies, "It's in my nature and I can't help it." Maybe that's how free will works. With apologies to the eliminative realists, it may be that our feeling of free will is, in fact, an illusion.

Determinism and free will are absolute concepts that are impossible to test directly. Therefore, their practical usefulness is limited and they are at best approximations that play a role within the framework of a specific model. The world itself, with all its stars, particles, and people, does what it does. The laws of nature are nothing more than our attempts to make models of the world. Our vantage point is limited and constantly evolving. There may be phenomena that will forever be beyond us. "In principle" is not enough; it is what is in practice that counts.

# Acknowledgments

ONE MORNING THERE WAS A KNOCK on my office door. A friendly man with a soft voice and intelligent eyes asked if he could come in. It was none other than Douglas Hofstadter, a guest at the Department of Physics and Astronomy for a few months, with whom I had already had several long and engaged discussions about life, the universe, and everything else. Now he wanted to make me aware of a problem during a lecture he had given at the university a few days earlier. The auditorium had been completely full. Before and after his lecture, enthusiastic readers of his international bestseller *Gödel, Escher, Bach* swarmed around him to get his signature in their well-thumbed copies. I had been in charge of the lecture and had done my best to introduce him adequately. This was what worried him. When he arrived onstage, there had been a problem with the sound and his grateful reply to me for the introduction had not been recorded. He wanted to tell me that he had met with the technical staff to make a new recording of the first minute. He had made sure that he was dressed in exactly the same way so that no one who watched the video would notice any difference. But this was not all. He suggested that we should once and for all decide this question: To what extent is consciousness an illusion? It was, of course, an offer I could not resist.

We had many long discussions—not only about consciousness and artificial intelligence but also about the nature of mathematics. Some of them form the basis of this book. I owe him my gratitude for all the inspiration and insights he conveyed to me. But I'm not so sure we agreed on the issue of consciousness. In fact, I'm not even sure that we really agreed on what we disagreed on.

Another person who has been an important inspiration to me is Max Tegmark. As I have explained, the fact that we completely disagree on crucial issues has been a prerequisite for the engaged discussions we have had on a couple of occasions—also in front of an audience. Others to whom I wish to express my great appreciation are Olle Häggström and Patrik Lindenfors, whom I have also very much enjoyed arguing with.

Anders Karlqvist, on the other hand, is one who shares my skepticism about how mathematical models are used, especially when it comes to living organisms. He read early drafts of *The World Itself* and made many valuable comments.

The person I have probably had the most discussions with is the playwright Erik Gedeon. Despite our different backgrounds, our views on the world itself have much in common.

I send a big thank-you to the book's Swedish editor, Emanuel Holm, who read all my arguments and helped me sharpen them. The weaknesses that remain are my own responsibility.

Thanks also to Fredrik Wikström, with whom I got lost many times, and my brother and literary critic Tommy Danielsson, who put me back me on the right track when I got too caught up in physics. Without the support and inspiration of my close family, Viktoria, Manne, and Olof, as well as my grown-up children, Oskar and Klara, there would not have been any book at all.

# Recommended Reading

HERE IS SOME RECOMMENDED READING for those who want to dig deeper into the various topics discussed in *The World Itself.*

### Everything Is Physics: An Introduction

For an overall critique of naïve materialism, I recommend *Are You an Illusion?* (Midgley 2014) and *Mind and Cosmos: Why the Materialist Neo-Darwinian Concept of Nature Is Almost Certainly False* (Nagel 2012). Thomas Nagel has in several works argued that there is something missing in our understanding of the world and points to a number of shortcomings of physical materialism. I share Midgley's and Nagel's critique but choose as an alternative to expand the definitions of matter and physics. An example of how to handle my conclusions on a personal level can be found in *This Life: Secular Faith and Spiritual Freedom* (Hägglund 2019). The philosophical direction of phenomenology plays an important role throughout *The World Itself.* I recommend introductions such as *Phenomenology: The Basics* (Zahavi 2018) and *Phenomenology* (Gallagher 2012). *The Phenomenon of Life: Toward a Philosophical Biology* (Jonas 1966) provides fascinating and eye-opening perspectives on the role of dualism in how we have viewed life throughout history.

### Living Beings Are Not Machines

A classic on the subject of life that is still worth reading is *What Is Life?* (Schrödinger 1944). An interesting modern summary of what one actually knows about biological life is *From Bacteria to Bach and Back: The Evolution of Minds* (Dennett 2017). At the same time, the book has a problematic bias that I oppose in several places in *The World Itself.*

### The Universe Is Not Mathematics

*Our Mathematical Universe* (Tegmark 2014) is in many ways a counterpoint to *The World Itself.* I have no objections to the strictly scientific content of Tegmark's book, but the existential conclusions are foreign to me. Much closer to my own view of mathematics is *Where Mathematics Comes From* (Lakoff and Núñez 2000), which shows how mathematics is rooted in bodily experiences. In *Incompleteness: The Proof and Paradox of Kurt Gödel* (Goldstein 2005), you can read about the man behind the theorems. As for the mathematical aspects, *Gödel's Proof* (Nagel and Newman 2001, revised edition) is a neat classic. *The Emperor's New Mind: Concerning Computers, Minds, and the Laws of Physics* (Penrose 1989) and *Shadows of the Mind: A Search for the Missing Science of Consciousness* (Penrose 1994) are fascinating reviews of Gödel's results and their consequences— even if some of the conclusions differ decisively from mine. A sober and factual review of what everything actually means can be found in "On the Philosophical Relevance of Gödel's Incompleteness Theorems," *Revue Internationale de Philosophie* 234, issue 4 (2005): 513–34. To those who want to immerse themselves in mathematical riddles on the theme of Gödel, I can recommend *Forever Undecided: A Puzzle Guide to Gödel* (Smullyan 1987). An inexhaustible source of inspiration worth returning to time and time again is *Gödel, Escher, Bach: An Eternal Golden Braid* (Hofstadter 1979).

### There Is a Difference Between Model and Reality

A book that has been useful for me is *Philosophy in the Flesh: The Embodied Mind and Its Challenge to Western Thought* (Johnson and Lakoff 1991). It analyzes in detail the Platonic delusion and offers the natural alternative.

### Computers Are Not Conscious

*Descartes' Bones: A Skeletal History of the Conflict Between Faith and Reason* (Shorto 2008) recounts the macabre and entertaining story of what happened to Descartes after death. *Descartes' Error: Emotion, Reason, and the Human Brain* (Damasio 1994) emphasizes that we do not only think with our brains but are dependent on the body as a whole. For diametrically opposed views of my own, I recommend *Life 3.0: Being Human in the Age of Artificial Intelligence* (Tegmark 2017) and *Superintelligence: Paths, Dangers, Strategies* (Bostrom 2014). *Consciousness Explained* (Dennett 1991) argues for a position in parts very far from my own. In *The World Itself,* I present myself as a physicalist. In *The Philosophy of Carl G. Hempel: Studies in Science, Explanation, and Rationality* (Fetzer 2001) one finds an interesting problematization of the concept. In the March 6, 1997, issue of *The New York Review of Books*, John Searle reviewed *The Conscious Mind: In Search of a Fundamental Theory* (Chalmers 1996). This resulted in a revealing exchange of opinions in *The New York Review of Books* May 15, 1997, issue.

### Not Everything Can Be Calculated

*Incomplete Nature: How Mind Emerged from Matter* (Deacon 2011) systematically explores the possibility that there are macroscopic laws of nature that cannot be derived from microscopic physics. *Agent Above, Atom Below: How Agents Causally Emerge from Their Underlying Microphysics* (Hoel

2018) analyzes similar thoughts in more detail. One of the most extraordinary and difficult books ever written is *Life Itself* (Rosen 1991). A bit more concise material can be found in *Essays on Life Itself* (Rosen 2000). A follow-up is *The Reflection of Life: Functional Entailment and Imminence in Relational Biology* (Louie 2013). One of the best overviews of these ways of looking at life is *Mind in Life: Biology, Phenomenology, and the Sciences of Mind* (Thompson 2007), while "Approaches to the Question, 'What is Life?': Reconciling Theoretical Biology with Philosophical Biology," *The Journal of Natural and Social Philosophy* 4, 1–2 (2008): 53–77, puts Rosen in a larger context.

### Man Is Not Unique

On the theme of other beings' inner worlds, "What Is It Like to Be a Bat?" *The Philosophical Review* 83, 4 (1974): 435–50), is a classic. These thoughts are further developed in *The View from Nowhere* (Nagel 1986) and *What Does It All Mean? A Very Short Introduction to Philosophy* (Nagel 1987). Some relatively modern and easily accessible books on nonhuman subjects are *Other Minds: The Octopus, the Sea, and the Deep Origins of Consciousness* (Godfrey-Smith 2016), *What a Fish Knows: The Inner Lives of Our Underwater Cousins* (Balcombe 2016), and *Brilliant Green:The Surprising History and Science of Plant Intelligence* (Mancuso and Viola 2015). A profound philosophical work with a surprising perspective is *The Primacy of Movement* (Sheets-Johnstone 2011, expanded second edition). It also provides an overview of relevant parts of phenomenology. Of course, you also have to get acquainted with *Nausea* (Sartre 1938).

### Does Free Will Exist?

*The Nonsense of Free Will: Facing Up to a False Belief* (Oerton 2012) is recommended to anyone who believes in free will.

The publication of this book is made possible by the support of the Brandt Jackson Foundation.

Bellevue Literary Press is devoted to publishing literary fiction and nonfiction at the intersection of the arts and sciences because we believe that science and the humanities are natural companions for understanding the human experience. We feature exceptional literature that explores the nature of consciousness, embodiment, and the underpinnings of the social contract. With each book we publish, our goal is to foster a rich, interdisciplinary dialogue that will forge new tools for thinking and engaging with the world.

To support our press and its mission, and for our full catalogue of published titles, please visit us at blpress.org.

Bellevue Literary Press
New York

3 9082 14621 1951